# Understanding Public Opinion Polls

# Understanding Public Opinion Polls

Jelke Bethlehem

CRC Press
Taylor & Francis Group
Boca Raton London New York

CRC Press is an imprint of the
Taylor & Francis Group, an **informa** business

CRC Press
Taylor & Francis Group
6000 Broken Sound Parkway NW, Suite 300
Boca Raton, FL 33487-2742

International Standard Book Number-13: 978-1-1380-6655-7 (Hardback)
International Standard Book Number-13: 978-1-4987-6974-7 (Paperback)

### Library of Congress Cataloging-in-Publication Data

Names: Bethlehem, Jelke G., author.
Title: Polling public opinion / Jelke Bethlehem.
Description: Boca Raton, FL : CRC Press, 2017.
Identifiers: LCCN 2017002246 | ISBN 9781498769747 (pbk.)
Subjects: LCSH: Public opinion polls.
Classification: LCC HM1236 .B48 2017 | DDC 303.3/8--dc23
LC record available at https://lccn.loc.gov/2017002246

**Visit the Taylor & Francis Web site at**
**http://www.taylorandfrancis.com**

**and the CRC Press Web site at**
**http://www.crcpress.com**

# Contents

# CHAPTER 1

# About Polls

## 1.1 THE WHEAT AND THE CHAFF

Are humans responsible for global warming? This was the only question in a poll of the English newspaper *The Telegraph*. It conducted the poll in 2013. The single-question poll, as shown in Figure 1.1, was conducted. It was filled in by 15,373 people. According to 30% of the respondents humans were completely responsible, 24% thought they were only partially responsible, and almost half of the respondents (46%) answered that global warming is a natural phenomenon.

The poll conducted as shown in Figure 1.1 is an example of a simple poll. It comprises only one question. Such polls are frequently encountered on websites. Often those are not very serious but sometimes are really intended to measure opinions on certain topics. It is better to ignore bad polls, but good polls can give more information about specific groups of people.

Polls are conducted every day all around the world. For each poll that looks interesting, it should always be checked whether its outcomes are valid. Does the poll really give a good idea of what is going on? The answer is sometimes "yes" and sometimes "no." There are many good polls. Their conclusions can be believed. There are, however, also many bad polls. It is better to ignore them. Unfortunately, it is not always easy to separate the wheat from the chaff.

Returning to the example of Figure 1.1, is the poll of *The Telegraph* a good one? A first impression can be obtained by looking at two aspects: the representativity of the sample and the design of the questionnaire.

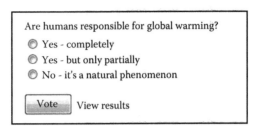

FIGURE 1.1 A single-question poll. (Adapted from a poll in *The Telegraph*, May, 2013.)

Looking at *representativity* means attempting to answer the question whether the sample is the representative for the population from which it came. More than 15,000 people participated in this poll. Are these respondents representative for all readers of *The Telegraph*? Or are there any substantial differences between the participants (the respondents) and those who did not participate? These questions can be answered only if it is clear how the sample was selected. The right way is to select a *random sample*, and the wrong way is to rely on some other forms of sampling, such as self-selection. Unfortunately, the poll of *The Telegraph* was based on self-selection: people spontaneously decided whether or not to participate in the poll. Self-selection usually leads to nonrepresentative samples, and therefore to wrong poll outcomes.

It is also important to look at the way in which the questions were asked in the poll. A researcher must ask each question in such a way that the respondents give a correct answer. This means that respondents must at least understand the questions. Furthermore, questions must be asked in a neutral fashion and must not lead respondents into a specific direction. In the case of the poll, it would have been better to include all answer options in the question text (*Are humans responsible for global warming, or is it a natural phenomenon?*). One may also wonder whether respondents understand terms like *global warming* and *natural phenomenon*. Another issue is why there is no option like *don't know*? What should people do who really do not know the answer?

Taking into account the probable lack of representativity of the sample and the poor design of the questionnaire, the conclusion can only be that the poll of *The Telegraph* was not a very good one. In addition, these are not the only aspects of a poll that should be checked. This is the reason why this book is written. The book helps us to assess the quality of polls. To do this, it introduces a *checklist for polls*. It consists of nine questions.

TABLE 1.1    The Checklist for Polls

|   | Question | Yes | No |
|---|----------|-----|----|
| 1 | Is there a research report explaining how the poll was set up and carried out? |  | √ |
| 2 | Is the poll commissioned or sponsored by an organization that has no interest in its outcomes? | √ |  |
| 3 | Is the target population of the poll clearly defined? | √ |  |
| 4 | Is a copy of the questionnaire included in the research report, or elsewhere available? | √ |  |
| 5 | Is the sample a random sample for which each person in the target population has a positive probability of selection? |  | √ |
| 6 | Are the initial (gross) sample size and the realized (net) sample size (number of respondents) reported? |  | √ |
| 7 | Is the response rate sufficiently high, say higher than 50%? |  | √ |
| 8 | Have the outcomes been corrected (by adjustment weighting) for selective nonresponse? |  | √ |
| 9 | Are the margins of error specified? |  | √ |

If the answers to all questions are "yes," the poll is probably OK, and it is safe to use its results. If the answer to one or more questions is "no," one should be careful. It may mean that there is not enough information to judge the quality of the poll, or it simply is a bad poll. Consequently, it is probably better to ignore it. Table 1.1 contains the nine questions of the checklist. Not everything may be clear. However, the book will explain all these aspects in more detail.

Table 1.1 shows how the poll of *The Telegraph* scored in the checklist. There are several no's. This is a clear signal that there is something wrong with the poll. Therefore, it is wise not to attach too much value to its outcomes.

## 1.2  WHAT IS A POLL?

A poll can be described as an instrument for collecting information about a group of people. The information is collected by means of asking questions to people in the group. This must be performed in a standardized way. Therefore, a questionnaire is used.

The researcher could ask every person in the group to fill in the questionnaire. Such a poll is called a *census* or a *complete enumeration*. Interviewing everybody, however, makes a poll expensive and time-consuming. Moreover, it would mean that everybody in the group is confronted with polls all the time. This increases the response burden. Consequently, motivation to participate drops, which may result in *nonresponse*. To avoid all these problems, a poll usually collects information on only a small part of the target population. This small part is the *sample.*

In principle, only information is obtained from the sampled persons. People, who are not in the sample, will not provide information. Fortunately, if the sample is selected in a *clever* way, conclusions can be drawn about the group as a whole. In this context, *clever* means that the sample is selected by means of *probability sampling.* This is a procedure that uses a form of random selection to determine which persons are selected, and which are not. This is called a *random sample.*

If it is clear how the selection mechanism works, and if the probabilities of being selected in the sample can be calculated, accurate estimates of characteristics of the target population can be computed. This means that the opinion of people in the sample can be used to accurately estimate the opinion in the whole population.

This is the *magic* of sampling: by selecting a sample of, say, 1000 people, the researcher can draw accurate conclusions about a group of many millions. The principle of sampling is not restricted to opinion polls. People do this kind of things often in daily life, for example, if they taste a spoon of soup, or they take a sip from a beer bottle. As the well-known Spanish writer Miguel de Cervantes (1547–1616) already said in his famous book *Don Quixote*: *By a small sample we may judge the whole piece.* Beware, however. This only works if the sample is selected by means of probability sampling (the soup must be stirred well). There are other sampling techniques like *quota sampling* and *self-selection*, which do not follow the fundamental principles of probability sampling. Therefore, there is no guarantee that these sampling approaches will allow the researcher to draw correct conclusions about the group.

There are *polls,* and there are *surveys.* There are no essential differences between polls and surveys. Those are both instruments to collect data from a sample by means of asking questions. There are, however, some practical differences. A poll is often small and quick. There are only a few questions (sometimes even only one) that have to be answered by a small sample of, say, 1000 people. A poll contains mainly opinion questions. A survey can have both opinion and factual questions. Analysis of the results will not require much time. Focus is on quickly obtaining an indication of the public opinion about a current issue. Surveys are often large and complex. They are typically conducted by large government organizations, such as national statistical institutes. More people will complete the questionnaire. The questionnaire may contain many questions and can have a complex structure. The analysis of the results may be time-consuming. The focus is more on precise and valid conclusions than on global indications. For all practical purposes, this book treats polls and surveys as similar.

## 1.3 CONDUCTING A POLL

At first sight, conducting a poll looks simple and straightforward. But there is more to it than one may think. Setting up and carrying out a poll is a process involving a number of steps and decisions. It is important to take a proper decision in each step. This does not always prove easy. Often decisions are a compromise between the costs of the survey and the quality of its outcomes. Nevertheless, one should always be aware of that wrong design decisions may lead to meaningless results. This section contains an overview of the steps to be taken in setting up a poll. The subsequent chapters treat many aspects in more detail.

### 1.3.1 The Target Population

The first step for the researcher is to decide which group of people must be investigated. This group is called the *target population*. The sample must be selected from this group, and, of course, the outcomes of the poll apply to this group, and no other group.

### 1.3.2 The Variables

What does the researcher want to know about the target population? Which properties of the people in the target population must be measured? Answering these questions comes down to selection of a set of relevant variables, also called the *target variables*. *Variables* are quantities that can assume any of the set of given values. The age of a person is a variable. It can have any value between 0 and, say, 120. Access to internet is also a variable (with two possible values "yes" and "no"). Likewise, opinion about an issue is a variable. The possible values of an opinion depend on the way it is measured. If the researcher asks whether someone is in favor or opposed, there are only two possible values (favor and oppose). If someone's opinion is asked by means of a five-point Likert scale, there are five possible values (strongly favor, somewhat favor, neither favor nor oppose, somewhat oppose, and strongly oppose).

### 1.3.3 The Population Characteristics

The collected data are used to draw conclusions about the status, opinion, or behavior of the target population as a whole. This usually comes down to computing a number of indicators. These indicators are called *population characteristics*. These population characteristics can be computed exactly if the individual values of the relevant target variables are known for every person in the target population. For example, if the population characteristic is the percentage of people having internet access, it can be computed if it is known for all people in the population if they have access to internet. If a poll is carried out, the values of target variables become available only for people in the sample. Therefore, population characteristics have to be *estimated*.

### 1.3.4 The Questionnaire

Data are collected by means of asking questions to people in the target population. Questions have to be asked in a consistent and objective way, so that comparable answers are obtained. Therefore, the questions must be asked to all respondents in exactly the same way. A *questionnaire* is the way to do this. Designing a questionnaire is a crucial aspect of a poll. Aspects like the format of questions (open, closed, check-all-that-apply, or numeric), the wording of the question texts, and the context of the questions all may have a substantial effect on the answers given. Errors in the questionnaire will lead to wrong answers to questions, and wrong answers will lead to wrong conclusions. If people in the target population speak different languages, it may be necessary to have different language versions of the questionnaire. Then, these versions must measure the same things in the same way. This is not always easy to realize.

### 1.3.5 The Mode of Data Collection

How to get the answers to the questions? How to collect the data? There are several ways to do this. These are called *modes of data collection*. One mode is to visit people at home and ask questions *face-to-face*. Another mode is calling people and asking questions by *telephone*. These two modes of data collection are *interviewer-assisted* modes because interviewers ask the questions. There are also *self-administered* modes. Then, respondents are on their own. There are no interviewers to help them. One example of a *self-administered* mode is sending the questionnaires by *mail* to people. Later, we will see that each mode has advantages and disadvantages.

Traditional polls use paper questionnaires. Nowadays, the computer can also be used to ask questions. This means the paper questionnaire is replaced by a digital one in the computer, laptop, tablet, or smartphone. This is called *computer-assisted interviewing* (CAI).

The quality of the data collected with CAI is usually better, and it takes less time to do a poll. For face-to-face interviewing, interviewers bring a laptop or tablet with them to the home of the people. This is *computer-assisted personal interviewing* (CAPI). For telephone interviewing, people are called from the call center of the poll organization. This is *computer-assisted telephone interviewing* (CATI).

The emergence of the internet and the World Wide Web rapidly made a new mode of data collection very popular: the *online poll*. Indeed, online data collection seems attractive as it is possible to collect a lot of data in an easy, cheap, and fast way. Unfortunately, this mode of data collection also has its problems. More details can be found in Chapter 7.

## 1.3.6 The Sampling Frame

Once a researcher decides how to collect the data (by selecting a mode of data collection), he or she must find a way to select a sample. A sampling frame is required for this. A *sampling frame* is a list of all people in the target population. This list must contain contact information. It must be clear for all persons in the list how to contact them. The choice of the sampling frame depends on the mode of data collection. For a face-to-face poll or a mail poll, addresses are required; for a telephone poll, telephone numbers are required; and for an online poll, the researcher preferably must have e-mail addresses.

It is important that the sampling frame exactly covers the target population. If this is not the case, specific groups may be excluded from the poll. This causes a lack of representativity. This affects the validity of the outcomes of your poll.

## 1.3.7 The Sampling Design

It is a fundamental principle of survey methodology that probability sampling must be applied to select a sample from the target population. Probability sampling makes it possible to draw valid conclusions about the target population as a whole. Moreover, it is possible to quantify the precision of estimates of population characteristics.

There are various ways of selecting a probability sample. The most straightforward one is selecting a *simple random sample* (with equal probabilities). Other sampling designs are selecting a *systematic sample, a stratified sample*, or a *two-stage sample*. On the one hand, the choice depends on practical aspects such as the availability of a sampling frame and the costs of data collection. On the other hand, it depends on how precise the estimates must be.

### 1.3.8 The Sample Size

How large must the sample be? There is no simple answer to this question. It can be shown that the precision of the estimates of population characteristics depends on the *sample size*. So, if very precise estimates are required, a large sample must be selected. If one is satisfied with less precision, a smaller sample may suffice. The sample size can be computed once it is known how precise the estimates must be.

### 1.3.9 Data Collection

If it is clear how the data will be selected (the mode of data collection), and a sample has been drawn from the target population, the fieldwork can start. This is the phase in which the researcher attempts to get the answers to the questions for all people in the sample. If it has been decided to do data collection face-to-face, interviewers must visit the sample persons. If the poll is conducted by telephone, interviewers must call the sample persons. For a mail poll, questionnaires must be sent to the sample persons by post. In the case of an online poll, the internet address of the questionnaire must be sent to the sample persons (by e-mail or by some other means).

All persons in the sample must complete the questionnaire. Unfortunately, often some questionnaires remain empty. For various reasons, people do not fill in the questionnaire. This is called *nonresponse*. There can be various causes: people cannot be contacted, they simply refuse to participate, or they are not able to answer the questions (e.g., due to a language problem). Nonresponse affects the validity of the poll results if it is selective. Unfortunately, nonresponse is often selective. Specific groups are under- or overrepresented in the poll.

### 1.3.10 Analysis

After data collection, the data must be analyzed. The first step in the analysis (and sometimes the only step) will often be estimation of various population characteristics, such as totals, means, and percentages. It may be informative to compute these estimates for several subpopulations into which the target population can be divided. For example, estimates can be computed separately for males and females, for various age groups, or for various regions in the country.

It should always be realized that estimates of population characteristics are computed, and not their exact values. This is because only sample data are available. Estimates have a *margin of error*. These margins of error must be computed and published, if only to avoid the impression that estimates represent the true values.

### 1.3.11 Nonresponse Correction

Almost every poll is affected by nonresponse. Nonresponse is often selective because specific groups are underrepresented in the poll. Selective nonresponse leads to wrong conclusions. To avoid this, a correction must be carried out. This correction is called *adjustment weighting*. This comes down to assigning weights to respondents. Respondents in overrepresented groups get smaller weights than respondents in underrepresented groups. This will improve representativity.

### 1.3.12 Publication

The final phase of the poll process is publication of the results. There should be a research report. Of course, this report will contain analysis results and conclusions. This must be done in such a way that readers of the report can understand it. Graphs are an import means of communication to do this. They help us to convey the message in the data.

The report should also describe how the poll was set up and carried out. Everybody must be able to check whether the principles for sound scientific survey research were applied. Among the things to report are a description of the target population, the sampling frame and sampling design, the sample size, the response rate, the correction for nonresponse, and the margins of error.

## 1.4 EXAMPLES OF POLLS

This chapter started with a very simple example of a poll. It was a poll with only one question. Many polls are much larger and much more complex than this one. Two examples illustrate this. The first example is the American Community Survey (ACS), and the second one is the Eurobarometer.

### 1.4.1 The American Community Survey

The ACS is conducted by the United States Census Bureau. The survey collects important information about the state of the United States. The collected information is used for making policy decisions at the federal and the state level. A detailed description of the ACS can be found on the website of the Census Bureau (http://www.census.gov/programs-surveys/acs/). Mervis (2015) also gives a good summary.

The questionnaire of the ACS contains 72 questions. The questions are about topics like ethnic background, level of educational, income, languages spoken, migration, disability, employment, and housing. These are typically factual questions and not opinion questions.

The ACS is a continuous survey. Each year a random sample of 3.5 million addresses is selected. The sample is divided into 12 monthly portions of 295,000 persons each. So, data collection is spread over the year. The sampling frame is the so-called master address file (MAF) that was constructed by the Census Bureau for the census in the year 2000. It is the official list of all addresses of houses in the United States.

The Census Bureau employs four different modes of data collection. It starts with a letter by mail that informs people at the selected addresses that they have to fill in the ACS questionnaire online. If there is no response after several days, sample addresses receive a postcard with a reminder and a paper version of the questionnaire. So, people now have a choice to complete the online questionnaire or the paper questionnaire. This leads to a response rate of between 50% and 55%.

The data-collection process does not stop here. In the case of nonresponse, the Census Bureau attempts to find telephone numbers of the nonresponding addresses. If these telephone numbers are found, interviewers try to complete the questionnaire in a telephone interview. This leads to an extra 10% response. To get rid of the remaining 35% nonresponse, face-to-face interviewing is used. Approximately 3000 interviewers go to the homes of the nonrespondents and ask them to fill in the survey questionnaire.

In the end, the response rate is over 95%. This is a very high response rate. This is caused mainly by the fact that the survey is mandatory. Respondents are legally obliged to answer all the questions, as accurately as they can. Failure to do so may result in a fine.

### 1.4.2 The Eurobarometer

The Eurobarometer is a system of opinion polls that have been conducted since 1974. Objective of the Eurobarometer polls is measuring the opinion of the citizens of the member states of the European Union about the process of integration. There are three types of Eurobarometer polls. The *Standard Eurobarometer* measures the opinion of the Europeans about integration, European institutes, and European policies. The polls also measure demographic and socioeconomic variables. Some variables are already measured for many years.

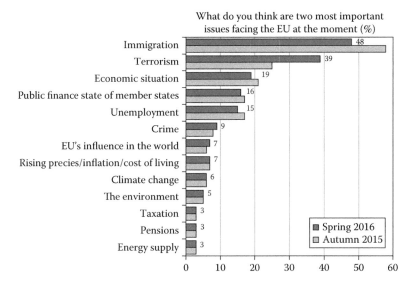

FIGURE 1.2 An example of a bar chart produced with Eurobarometer data. (From Eurobarometer, GESIS, http://www.gesis.org/eurobarometer-data-service.)

The Standard Eurobarometer can therefore be used for longitudinal research. The poll is conducted twice a year.

Figure 1.2 presents an example. The bar chart was constructed using data from two consecutive polls that were conducted in the autumn of 2015 and the spring of 2016. It shows that immigration and terrorism are considered the two most important issues.

There is also the Special Eurobarometer. It measures opinions about special themes. The questions of the Special Eurobarometer are usually embedded in the Standard Eurobarometer. The third type of Eurobarometer is the Flash Eurobarometer. Such a poll is used for quickly measuring the opinion of the European citizens on a specific topic. The questionnaire usually contains only a few questions.

The Directorate-General for Communication (DG COMM) of the European Commission is responsible for the Eurobarometer polls. It decides which questions will be included in the polls, and which outcomes are published. DG COMM is not very transparent with respect to giving insight into the methodological aspects of the polls. For example, no response rates are made available. As they are considered technical information, they are not published. This makes it difficult to assess the quality of the outcomes of the polls.

There are two Standard Eurobarometer polls each year. As an example, the spring poll in 2015 is described in some more detail. The target population of this poll not only consisted of the citizens of the 28 member states but also the citizens of five candidate countries.

The questionnaire contained 68 questions. The questions were on topics like the future of Europe, European citizenship, the financial and economic crisis, the current and future financial situation of people, quality of life, internet use, and various demographic aspects. Many different languages are spoken in the countries of the European Union. Consequently, there were 44 different versions of the questionnaire. Even for a very small country like Luxembourg, there were three different versions (in German, French, and Luxembourgish).

The sample was selected using a rather complex sampling procedure. To obtain good-quality data, it was decided to apply face-to-face interviewing. So, a random sample of addresses was required. To that end, each country was divided into a number of geographical areas (at the so-called NUTS2 level). In each geographical area, a number of small geographical units (cluster of addresses) were randomly selected. In each cluster, a random starting address was selected.

Further addresses were selected by a random route procedure. This comes down to taking a walk from the starting address and knocking at the door of each $n$th address. At each selected address, a random respondent was drawn using the first birthday rule, that is, the person was elected with his or her closest birthday. If no contact was possible at an address, two recall attempts were made. The interviewers continued their walk until they had completed their quota for a cluster of addresses.

In each country, a sample of at least 1000 persons had to be selected. In a few very small countries, this was reduced to 500 persons. Sampling and interviewing was continued until the prescribed sample size was reached. This creates the impression that there is no nonresponse. On the contrary, there is nonresponse, but it is hidden and not reported. So, it is difficult to assess the quality of the survey outcomes.

## 1.5 SUMMARY

Polls and surveys are instruments for collecting data about a specific target population. Measurement comes down to asking questions. Not all people in the target population are investigated, but only a sample of them.

To be able to generalize the results of the analysis to the sample data to the target population, the sample must be selected properly. *Properly* means here that the sample must be selected by means of some form of probability sampling, where each individual in the target population has a positive probability of selection.

There are many polls, particularly during election campaigns. Not all polls are good. There are also many bad polls. To be able to assess the quality of a poll and its results, it is good to know how the poll was conducted. This calls for a research report that not only contains the conclusions drawn from it but also explains how the poll was set up and carried out.

This chapter started with a very simple example of a poll. It was a poll with only one question. Many polls are much larger and much more complex than this one. Two examples illustrate this. The first example is the ACS and the second one is the Eurobarometer.

This book is intended for people who are in some way confronted with polls. It may be due to the fact that they have to judge the outcomes of polls as a part of their professional work. One can think of journalists, decision makers, political scientists, and politicians. This book helps them to establish whether a poll is good or bad, and whether they safely can use the poll results, or whether it is better to ignore the poll.

This book can also be a useful source of information for those who want to carry out a poll themselves and are not an expert in this area. They can use this book as a guideline. Of course, this book is a not-so-technical introduction in the world of opinion polling.

# CHAPTER 2

# Some History

## 2.1 THE ORIGINS OF STATISTICAL DATA COLLECTION

All through history, there has always been a need for statistical overviews about the state of countries. So, there have always been some forms of data collection. There were already agricultural censuses in Babylonian times. These were conducted shortly after the invention of the art of writing. Ancient China counted its people in order to determine the revenues and the military strength of its provinces. There are also accounts of statistical overviews compiled by Egyptian rulers long before the birth of Christ. The rulers of the Rome Empire conducted censuses of people and property. They used the data to establish the political status of citizens and assess the military strength and tax obligations to the state. An example is the numbering of the people of Israel, leading to the birth of Jesus in the small town of Bethlehem (see Figure 2.1).

Opinion polls can be traced back to ancient Greece. Around 500 BC, many city–states had a form of government based on democratic principles. All free native adult males could express their opinion on topics such as declaring war, dispatching diplomatic missions, and ratifying treaties. People turned opinions into decisions by attending popular assemblies and voting on these issues. Note that every member of the target population gave his opinion. So one could argue this was more a referendum than an opinion poll.

For a long period in history, data collection was based on complete enumeration of the target population. Every person in the target population had to provide information. The important idea of sampling emerged only at the end of the nineteenth century. It had taken many years before this idea was accepted.

FIGURE 2.1    Census in Bethlehem (Pieter Brueghel, 1605–1610).

This chapter gives a global overview of historical developments of data collection from the use of censuses, via the rise of survey sampling, to online polls.

## 2.2 THE CENSUS ERA

Ancient empires (China, Egypt, and Rome) collected statistical information mainly to find out how much tax they could collect and how large their armies could be. Data collection was rare in the Middle Ages. A good example was the census of England ordered by William, the Conqueror, in 1086. He became the King of England after he conquered it from Normandy in 1066. He wanted an overview of the land and resources being owned in England at the time, and how much taxes he could raise. All information was collected in the *Domesday Book* (see Figure 2.2). The compilation of the book started in the year 1086 AD. The book recorded information about each manor and village in the country. There was information about more than 13,000 places, and on each county, there were more than 10,000 facts. To collect all these data, the country was divided into regions, and in each region, a group of commissioners was appointed from among the greater lords. Each county within a region was dealt with separately. There

FIGURE 2.2   The *Domesday Book*.

were sessions in each county town. The commissioners summoned all those required to appear before them. They had prepared a standard list of questions. For example, there were questions about the owner of the manor, the number of free man and slaves, the area of woodland, pasture and meadow, the number of mills and fishponds, to the total value of the property, and the prospects of getting more profit. The *Domesday Book* still exists, and county data files are available online.

Another interesting example of compilation of statistics can be found in the Inca Empire that existed between 1000 and 1500 AD in South America. Each Inca tribe had its own statistician, called the *Quipucamayoc*. This man kept records of, for example, the number of people, the number of houses, the number of llamas, the number of marriages, and the number of young men that could be recruited for the army. All these facts were recorded on a *quipu*, a system of knots in colored ropes, see Figure 2.3.

At regular intervals, couriers brought the quipus to Cusco, the capital of the kingdom, where all regional statistics were combined into national statistics. The system of Quipucamayocs and their quipus worked remarkably well. Unfortunately, the system vanished with the fall of the empire.

An early census took place in Canada in 1666. Jean Talon, the intendant of New France (as Canada was called), ordered an official census

FIGURE 2.3   The Quipucamayoc, the Inca-statistician. (Reprinted with permission of ThiemeMeulenhoff from Burland 1971.)

of the French colony to measure the increase in population since its foundation in 1608. Name, age, sex, marital status, and occupation of every person were recorded. New France turned out to have a population of 3,215 persons. The first census in the United States was held in 1790. The population consisted of 3,929,214 people at that time.

The first censuses in Europe were undertaken by the Scandinavian countries: The census in Sweden–Finland took place in 1746. A census had already been proposed earlier, but the initiative was rejected because of a religious reason: *it corresponded to the attempt of King David who wanted to count his people.*

The industrial revolution in the nineteenth century was an important era in the history of statistics. It brought about substantial changes in society, science, and technology. Urbanization, industrialization, democratization, and emerging social movements created new demands for statistical information. This was a period in which the foundations of many principles of modern social statistics were laid. Several central statistical bureaus, statistical societies, statistical conferences, and statistical journals emerged as a result of these developments (see Figure 2.4).

FIGURE 2.4    A census-taker interviews a family in 1870. (Library of Congress, LC-USZ62-93675.)

## 2.3  THE RISE OF SURVEY SAMPLING

The development of modern sampling theory started around the year 1895. This was the year in which Anders Kiaer, the founder and first director of Statistics Norway, published his *Representative Method* (Kiaer, 1895). He described it as an inquiry in which a large selection of persons was questioned. This selection should be a *miniature* of the population. He selected people in such a way that groups were present in the right proportions. This would now be called *quota sampling*. The proportions were obtained from previous investigations. Anders Kiaer stressed the importance of *representativity*. His argument was that, if a sample was representative with respect to some variables (with a known population distribution), it would also be representative with respect to the other variables in the survey.

A basic problem of the Representative Method was that there was no way of establishing the accuracy of estimates. The method lacked a formal theory of inference. It was Arthur Bowley, who made the first steps in this direction in 1906. Bowley (1906) showed that for large samples, selected at random from the population, the estimate had an approximately normal distribution. Therefore, he could compute the variance of estimates and use them as an indicator of their precision.

From this moment on, there were two methods of sample selection. The first one was Kiaer's Representative Method, based on *purposive selection* (*quota sampling*), in which representativity played a crucial role, and for which no measure of the precision of the estimates could be obtained. The second one was Bowley's approach, based on *random sampling*, and for which an indication of the precision of estimates could be computed. Both methods existed side by side for a number of years. This situation lasted until 1934, in which year the Polish scientist Jerzy Neyman published his now famous paper. Neyman (1934) developed a new theory based on the concept of the *confidence interval*. The confidence interval is still used nowadays to describe the *margin of error* of an estimate.

Jerzy Neyman did more than inventing the confidence interval. Based on an empirical evaluation of Italian census data, he could prove that the Representative Method failed to provide satisfactory estimates of population characteristics. As a result, the method of purposive sampling fell into disrepute.

The classical theory of survey sampling was more or less completed in 1952 when Horvitz and Thompson (1952) published their general theory for constructing unbiased (valid) estimates. They showed that, whatever the selection probabilities are, as long as they are known and positive, it is always possible to construct valid estimates. Since then, probability sampling has become the preferred sample selection method. It is the only general sampling approach with which valid inference can be made about a population using sample data.

## 2.4 OPINION POLLS

You can see opinion polls as surveys that measure attitudes or opinions of a group of people on political, economic, or social topics. The history of opinion polls in the United States goes back to 1824. In this year, two newspapers, *The Harrisburg Pennsylvanian* and *The Raleigh Star*, attempted to determine political preferences of voters prior to the presidential election of that year. These early polls did not pay much attention to sampling aspects. Therefore, it was difficult to establish the accuracy of their results. Such opinion polls were often called *straw polls*. This expression goes back to rural United States. Farmers would throw a handful of straws into the air to see which way the wind was blowing. In the 1820s, newspapers began doing straw polls in the streets to see how political winds blew.

It took until the 1920s before more attention was paid to sampling aspects. At that time, Archibald Crossley developed new techniques for measuring American public's radio listening habits. Moreover, George Gallup worked out new ways to assess reader interest in newspaper articles. Lienhard (2003) describes these developments. Gallup applied a sampling technique called *quota sampling*. The idea was to investigate groups of people who were representative for the target population. Gallup sent out hundreds of interviewers across the country. Each interviewer was given quota for different types of respondents: so many middle-class urban women, so many lower class rural men, and so on.

The presidential election of 1936 turned out to be decisive for the way in which opinion polls were conducted. This is described by Utts (1999). At the time, the *Literary Digest Magazine* was the leading polling organization. This magazine conducted regular *America Speaks* polls. The sample consisted of addresses obtained from telephone directory books and automobile registration lists. The sample size for the election poll was very large: 2.4 million people.

Gallup's approach was in great contrast with that of the *Literary Digest Magazine*. For his quota sample he carried out *only* 50,000 interviews. Gallup correctly predicted Franklin Roosevelt to be the new president of the United States, whereas Literary Digest predicted that Alf Landon would beat Franklin Roosevelt. The results are summarized in Table 2.1.

How could a prediction based on such a large sample be so wrong? The explanation was a fatal error in the sampling procedure of *Literary Digest Magazine*. The automobile registration lists and telephone directories were not representative samples. In the 1930s, cars and telephones were typically owned by the middle and upper classes. More well-to-do Americans tended to vote Republican, and the less well-to-do were inclined to vote Democrat. Therefore, Republicans were systematically overrepresented in the Literary Digest sample.

As a result of this historic mistake, *Literary Digest Magazine* ceased publication in 1937. And opinion researchers learned that they should rely on more scientific ways of sample selection. They also learned that the way a sample is selected is more important than the size of the sample.

TABLE 2.1  The Presidential Election in the United States in 1936

| Candidate | Prediction by Literary Digest (%) | Prediction by Gallup (%) | Election Result (%) |
|---|---|---|---|
| Roosevelt (D) | 43 | 56 | 61 |
| Landon (R) | 57 | 44 | 37 |

TABLE 2.2   The Presidential Election in the United States in 1948

| Candidate | Prediction by Gallup (%) | Election Result (%) |
|---|---|---|
| Truman (D) | 44 | 50 |
| Dewey (R) | 50 | 45 |

Gallup's quota-sampling approach turned out to work better than Literary Digest's haphazard-selection approach. However, Jerzy Neyman had shown already in 1934 that quota sampling can lead to invalid estimates too. Gallup was confronted with the problems of quota sampling in the campaign for the presidential election of 1948. Harry Truman was the Democratic candidate, and Thomas Dewey was the Republican one. Table 2.2 summarizes Gallup's prediction and the real election result.

The sample of Gallup's poll consisted of 3250 people. Gallup concluded from the poll that Thomas Dewey would win the election. Some newspapers were so convinced of Dewey's victory that they already declared him the winner of the elections in their early editions. This prediction turned out to be completely wrong, however (see Figure 2.5).

Gallup predicted that Dewey would get 50% of the votes. That was 5% more than the election result (45%). An analysis of the polls of 1936 and 1948 showed that the quota samples contained too many Republicans and hence too few Democrats. This did not lead to a wrong prediction in 1936

FIGURE 2.5   Truman showing a newspaper with a wrong prediction.

because the real difference between the two candidates was much more than 5%. The difference between the two candidates was smaller in 1948. Therefore, the lack of representativity in Gallup's sample led to a wrong prediction.

These examples show the dangers of quota sampling. Quota samples are not based on random selection. Instead, interviewers are instructed to select groups of people in the right proportions. However, this can be done only for a limited number of variables, such as gender, age, level of education, and race. Making a sample representative, with respect to these variables, does not automatically guarantee representativity with respect to other variables, like voting behavior. The best guarantee for representativity is applying some form of random sampling. The most straightforward way of doing this is giving each person the same probability of selection. This is called a simple random sample.

## 2.5 FROM TRADITIONAL TO COMPUTER-ASSISTED INTERVIEWING

The traditional way of carrying out a poll is collecting data with a paper questionnaire. Respondents or interviewers fill in these forms, after which the data are entered in the computer. Then, the data can be analyzed. Traditional polls come in three forms: face-to-face, telephone, and mail.

A *face-to-face poll* is the most expensive of the three modes of data collection. Interviewers contact people in the sample face-to-face. The interviewers can visit them at home, or approach them on the street, or in a shopping mall. Well-trained interviewers persuade people who are reluctant to participate in the poll. Therefore, the response rate of a face-to-face poll is usually higher than that of a telephone or mail poll.

The interviewers can also assist respondents in finding the right answers to the questions. This often leads to better data. It should be noted, however, that the presence of interviewers can also be a drawback. Interviewers should not *push* respondents in a specific direction. Moreover, respondents are more inclined to give a true answer to a sensitive question if there are no interviewers around. If there is an interviewer present, a respondent may give socially desirable answer.

It will be clear that it is of crucial importance to have well-trained and experienced interviewers. Insufficiently trained and less-experienced interviewers can cause a lot of damage, leading to unanswered or incorrectly answered questions.

The high quality of face-to-face polls comes at a price. Interviewers have to travel around from one sample person to another. So, they can do only a limited number of interviews per day. This makes a face-to-face poll a costly and time-consuming process.

A *telephone poll* is a somewhat less costly form of data collection. Interviewers are also required for this mode of data collection, but not as many as for face-to-face interviewing. Telephone interviewers call people from a call center. They do not lose time traveling from one respondent to the next. Therefore, they can conduct more interviews in the same amount of time. This makes a telephone poll less expensive.

Telephone interviewing has some drawbacks. Interviews cannot last too long, and questions may not be too complicated. Another complication is the lack of a proper sampling frame. Telephone directories are usually incomplete. Many telephone numbers are missing. This includes part of the landline telephone numbers and almost all mobile phone numbers. Therefore, telephone directories are poor sampling frames.

A way to avoid the undercoverage problems of telephone directories is to apply *random digital dialing* (RDD). This means a computer algorithm is used for computing random telephone numbers. Random digital dialing has the drawback that response rates are often very low. This is caused by the increased use of answering machines and caller ID (a telephone service that makes the telephone number of the caller visible to the called person), and an increase of unsolicited telephone calls, including many polls and calls from telemarketers (which sometimes attempt to mimic polls).

A *mail poll* is even less expensive. Paper questionnaires are sent by ordinary mail to the people selected in the sample. They are asked to complete the questionnaire and send it back. A mail poll does not involve interviewers. Therefore, it is a cheaper mode of data collection than face-to-face or telephone poll.

The absence of interviewer also has disadvantages. They cannot help respondents to answer questions. This may reduce the quality of the collected data. Mail polls also put high demands on the design of the questionnaire. It should be clear to all respondents how to navigate through the questionnaire and how to answer questions.

As the persuasive power of interviewers is absent, response rates of mail polls are low. Of course, reminders can be sent, but this is often not very successful. More often, questionnaires and other poll documents end up in the pile of old newspapers.

Traditional modes of data collection are slow and error-prone. The completed questionnaires often contain errors. Therefore, the collected data must always be checked, and detected errors must be corrected. Fortunately, the advent of microcomputers in the last decades of the twentieth century made it possible to use microcomputers for data collection. Thus, *computer-assisted interviewing* (CAI) was born. The paper questionnaire was replaced by an electronic one. Interviewing software guided the interviewers or respondents through the questionnaire and checked the answers. CAI has three major advantages: (1) It simplifies the work of the interviewers, because they are no longer in charge of following the correct route through the questionnaire; (2) it improves the quality of the collected data, because answers can be checked and corrected during the interview; and (3) it considerably reduces time needed to process the data. Thus, a poll can be done quicker, the data are better, and it is less expensive. More about the benefits of CAI can be found, for example, in Couper et al. (1998).

CAI comes in three modes. These modes are the electronic analogs of the traditional face-to-face, telephone, and mail polls. The first mode of CAI emerged in the 1960s. It was *computer-assisted telephone interviewing* (CATI). The first nationwide telephone facility for polls was established in the United States in 1966. CATI-interviewers operated a computer running the interview software. When instructed so by their computer, they attempted to contact a person by telephone. If this was successful, and this person agreed to participate in the poll, the interviewer started the interviewing program. The first question appeared on the screen. If this question was answered correctly, the software proceeded to the next question on the route through the questionnaire.

Many CATI systems have a tool for *call management*. Its main function is to offer the right phone number at the right moment to the right interviewer. This is particularly important if an interviewer has made an appointment with a respondent for a specific time and date. Such call-management systems also deal with special situations such as a busy number (try again after a short while) or no answer (try again later). This all helps us to increase the response rate.

The next mode of CAI that emerged was *computer-assisted personal interviewing* (CAPI). It stands for face-to-face interviewing in which the paper questionnaire is replaced by an interviewing program on a portable computer. CAPI emerged in the 1980s when lightweight laptop computers made face-to-face interviewing with a computer feasible.

The first experiments with CAPI were carried out in 1984. It became clear that laptops could be used for interviewing provided they were not too heavy. Moreover, the readability of the screen should always be sufficient, even in bad conditions like a sunny room. And the battery capacity should be sufficient to allow for a day of interviewing without recharging.

Regular use of CAPI started in 1987 when Statistics Netherlands deployed portables for its Labor Force Survey. Approximately 400 interviewers were equipped with Epson PX-4 laptop computers (see Figure 2.6). After a day of work, they returned home and connected their computer to the power supply to recharge the batteries. They also connected their laptop to a telephone and modem. At night, when the interviewers were at sleep, their computers automatically called Statistics Netherlands and uploaded to collected data. New sample addresses were downloaded in the same session. In the morning, the computer was ready for a new day of interviewing.

The computer-assisted mode of mail interviewing also emerged in the 1980s. It was called *computer-assisted self-interviewing* (CASI), or also *computer-assisted self-administered questionnaires* (CASAQ). The electronic questionnaire was sent to the respondents. They completed the questionnaire on their computer and sent the data back to the poll organization. Early CASI applications used diskettes or a telephone and modem to send the questionnaire, but nowadays, it is common practice to download it from the internet.

FIGURE 2.6   The Epson PX-4 portable was used for the first CAPI-poll. (Reprinted with permission of John Wiley & Sons from Bethlehem 2009.)

A CASI poll is only meaningful if all respondents have a computer on which they can run the interview program. Initially, this was not the case for households. Therefore, use of CASI in household polls was not widespread in the early days. CASI for business surveys was more successful. One early example was the production of fire statistics by Statistics Netherlands. It started collecting data for these statistics by means of CASI already in the 1980s. Diskettes were sent to the fire brigades. They ran the questionnaire on their MS-DOS computers. The answers were stored on the diskette. After having completed the questionnaire, they returned the diskette to Statistics Netherlands.

## 2.6 ONLINE POLLS

The rapid development of the internet in the last decades has led to a new mode of data collection: the online poll. This mode is sometimes also called computer-assisted web interviewing (CAWI).

The development of the internet started in the early 1970s when the U.S. Department of Defense decided to connect computers of research institutes. Thus, ARPANET was born. It became a public network in 1972. Software was developed to send messages over the network. So e-mails emerged.

Researchers realized that they could use e-mail to conduct polls. An early experiment in 1983 described by Kiesler and Sproull (1986) showed that the costs of e-mail polls were much less than those of mail polls. Response rates were more or less similar. And the turnaround time of the e-mail polls were much shorter.

E-mail polls also had drawbacks. They have almost no layout facilities. So it was impossible to give an e-mail message the look and feel of a questionnaire. Nowadays, e-mail systems have more layout facilities, but they are still too limited. Another drawback of e-mail polls was that it was impossible to control navigation through the questionnaire. And certainly in the first years, there were severe coverage problems, as the percentage of people with internet was still low. As a result of these problems, online polls were not a big success.

This all changed in 1995, when HTML 2.0 became available. HTML (HyperText Markup Language) is a markup language for web pages. The first version of HTML was developed by Tim Berners Lee in 1991 and thus the World Wide Web emerged. A strong point of HTML 2.0 was the possibility to have data entry forms on a computer screen. Computer users could enter data, and these data were sent from the computer of the user to the server of the researcher.

HTML 2.0 made it possible to design questionnaire forms and to offer these forms to respondents. And so web polls emerged. These online polls are almost always self-administered: respondents visit the website and complete the questionnaire by answering the questions.

Not surprisingly, researchers started using online polls, as they seem to have some attractive advantages:

- Nowadays, many people are connected to the internet. So an online poll has become a simple means to get access to a large group of potential respondents.

- Poll questionnaires can be distributed at very low costs. No interviewers are needed, and there are no printing or mailing costs.

- An online poll can be launched very quickly. Little time is lost between the moment the questionnaire is ready and the start of the fieldwork.

- Online polls offer new, attractive possibilities. For example, you can use multimedia (sound, pictures, animation, and movies) in the questionnaire.

Online polls became quickly very popular. In countries with high internet coverage (like The Netherlands with a coverage of 96%), almost all polls are now online polls. For example, during the campaign for the parliamentary election in September 2012, the four main poll organizations in The Netherlands were Maurice de Hond, Ipsos, TNS NIPO, and GfK Intomart. All their polls were online polls. Therefore, they could do polls very quickly. Some even conducted a poll every day.

So online polls seem to be a fast, cheap, an attractive means of collecting large amounts of data. There are, however, methodological issues. One of these issues is sample selection. How to draw a simple random sample for an online poll? It would be ideal to have a sampling frame containing e-mail addresses for all people in the target population. Unfortunately, such a sampling frame is almost never available. A way out could be to use an address list and send letters with a link and a unique login code to a sample of addresses. Unfortunately, this makes an online poll less easy, less fast, and less cheap.

Not every new development in data collection is an improvement. Regrettably, many researchers bypass nowadays the fundamental principles of probability sampling. They do not draw a random sample but rely on self-selection. They simply put the questionnaire on the web. Respondents are people who have internet, accidently see an invitation to participate in the poll, spontaneously go to online questionnaire, and answer the questions. Self-selection samples are usually not representative as they particularly contain people who like to do polls on the internet and are interested in the topic of the poll.

If self-selection is applied, the selection probabilities are unknown. Therefore, it is not possible to compute unbiased estimates of population characteristics. It is also not possible to compute confidence intervals or margins of errors. As a result, the sample outcomes cannot be generalized to the target population. There are also other problems with self-selection polls. One is that also people from outside the target population can participate. They *pollute* the poll. Another problem is that groups of people may attempt to manipulate poll results. It will be clear that it is wise to avoid online self-selection polls as much as possible.

## 2.7 SUMMARY

This chapter gave an overview of the every changing landscape of statistical data collection. It all started a long time ago with rulers of countries and empires who had a need of statistical overviews. For a long time, these statistics were based on complete enumeration of the population.

It took until the first ideas about sampling emerged in the nineteenth century. Ander Kiaer proposed his Representative Method in 1895. His method resembled quota sampling. He could not prove that his sample outcomes were accurate estimates.

The breakthrough was the proposal of Arthur Bowley in 1906 to draw a probability sample. Then probability theory can be used to prove that estimates have a normal distribution. Thus, valid estimates of population characteristics can be obtained. Moreover, margins of error of the estimates can be computed.

The Representative Method and probability sampling existed side by side for many years. The discussion about both approaches lasted until 1934, when Jerzy Neyman showed that probability sampling was superior to quota sampling.

The computer was used more and more for data collection in the second half of the twentieth century. CAI merged. This had no impact on the methodological framework, but it made conducting a poll more efficient and improved the quality of the collected data.

The advent of the internet made it possible to carry out polls online. In a short period of time, they became very popular. At first sight, they seem to be a simple, cheap, and fast way to collect a lot of data. Unfortunately, many online polls rely on self-selection instead of probability sampling. From a methodological perspective, this is a step back.

Given the popularity of online polls on the hand, and the methodological challenges on the other, one chapter is devoted completely to online polls, that is, Chapter 8.

# CHAPTER 3

# The Questionnaire

## 3.1 ASKING QUESTIONS

The questionnaire is an important component of a poll. It is a measuring instrument that is used for collecting and recording data. Unfortunately, it is not perfect. Someone's length can be measured with a measuring staff, and someone's weight can be determined with a weighing scale. These physical measuring devices are generally very accurate. The situation is different for a questionnaire. It only indirectly measures someone's opinion or behavior. It has been shown many times that even the smallest of changes in a question text can lead to major changes in the answers. If a wrong question is asked, a different concept is measured, or a wrong answer is recorded. If answers are wrong, estimates of population characteristics are wrong. And if the estimates are wrong, wrong conclusions are drawn.

Publications about the results of a poll are not always clear about how the data were collected. Probably, a questionnaire was used, but often this questionnaire is not included in the released information about the poll. So, it is not possible to determine whether the questionnaire was good or bad. Still, it is important to take a look at the questionnaire. If the questionnaire is OK, the answers to the questions can be trusted. If there are all kinds of problems in the questionnaire, it may be better to ignore poll results. If it is not possible to take a look at the questionnaire, one should be cautious.

Some polling organizations do not publish the questionnaire, but others are very transparent. An example is the *Eurobarometer*. The Eurobarometer is an opinion poll that is conducted regularly in the member states of the European Union. The European Commission is responsible for the

Eurobarometer. The poll documentation is published on the website of GESIS. It is also denoted by Leibniz-Institute for the Social Sciences. It is the largest infrastructure institution for the social sciences in Germany. See www.gesis.org/en/eurobarometer-data-service/home/. For each of the 28 EU countries, there is a questionnaire, and some countries have even more than one questionnaire, like Belgium (French and Flemish), Finland (Finnish and Swedish), and Luxembourg (German, French, and Luxembourgish). All questionnaires can be downloaded from the GESIS website.

An example makes clear how sensitive question texts are. The Conservative Party in Britain was the winner of the general election in May 2015. Hence, party leader David Cameron had to keep his promise to hold a referendum on Britain's membership of the European Union by the end of 2017. The prospect of this referendum generated a lot of discussion, and of course there were many polls measuring the opinion of the British about leaving the European Union. One of these polls was conducted in December 2015 by the market research company Ipsos MORI. It was a telephone poll, and the sample size was 1040 people. The poll contained a question about reducing the voting age from 18 to 16 years for the referendum. There were two different versions of this question, which are as follows:

- Do you support or oppose reducing the voting age from 18 to 16 in the referendum on Britain's membership of the European Union?

- Do you support or oppose giving 16- and 17-year olds the right to vote in the referendum on Britain's membership of the European Union?

Both questions asked essentially the same thing. Only the text was different. The questions were randomly assigned to the respondents. In the end, 536 respondents answered the first question, and 504 respondents the second one. Table 3.1 shows the results.

There are substantial differences between the two versions. If the two categories *strongly support* and *tend to support* are combined into one category *support*, and also *strongly oppose* and *tend to oppose* are combined into *oppose*, then for the first version of the question, a majority of 56% oppose reducing the voting age. Only 37% support it. For the second version of the question, the situation is completely different. Only a minority of 41%

TABLE 3.1  Answers to Different Versions of a Question

| | Reducing the Voting Age from 18 to 16 Years (%) | Giving 16- and 17-Year Olds the Right to Vote (%) |
|---|---|---|
| Strongly support | 20 | 32 |
| Tend to support | 17 | 20 |
| Neither support nor oppose | 6 | 5 |
| Tend to oppose | 18 | 14 |
| Strongly oppose | 38 | 27 |
| Don't know | 1 | 1 |
| Total | 100 | 100 |

*Source:* Ipsos MORI (2015).

opposes reducing the voting age, whereas a majority of 52% supports it. So, one may wonder whether the two versions really measure the same thing. This example shows how crucial the correct wording of the question is.

Indeed, asking questions is a delicate process in which a lot can go wrong. Therefore, it is wise to be careful when using the results of a poll. It is a good idea to take a good look at the questionnaire first. And if the questionnaire is not available, be aware that the collected data may contain errors. If a new questionnaire is designed, it is also important to pay attention to aspects like the definition of the questions, and the order of the questions. Moreover, questionnaires should be tested.

At first sight, asking questions and recording answers seem so simple. But it is more complex than one might think. Schwarz et al. (2008) describe the tasks involved in answering a poll question as follows:

1. Respondents need to understand the question to determine which information they should provide.

2. Respondents need to recall relevant information from memory. In the case of an opinion question, they may not have a ready-for-use answer stored in their memory. Instead, they have to form a judgment on the spot, based on whatever relevant information comes to mind at the time.

3. Respondents need to report their answer. They can rarely do it in their own words. They usually have to reformat their answer in order to fit one of the response alternatives provided by the questionnaire.

4. Respondents need to decide whether to report the true answer or to report a different answer. If a question is about a sensitive topic, they may decide to refuse giving an answer. If an answer is socially undesirable, they may change their answer into a more socially desirable one.

This chapter will show that a lot can go wrong in the process of asking and answering questions. Therefore, it is import to test a questionnaire. It is often not clear from the documentation whether the questionnaire was really tested, and how it was tested. Anyway, it is always useful to take a look at the questions. This chapter lists a number of aspects that can be checked. The results of these checks should help us to judge the questionnaire.

## 3.2 ASKING FACTUAL AND NONFACTUAL QUESTIONS

When inspecting the questions in the questionnaire, it is good to follow Kalton and Schuman (1982) and to distinguish factual and nonfactual questions. *Factual questions* ask about facts and behavior. There is always an individual *true value*. Often, it is possible to obtain this true value, at least in theory, by some other means. Examples of factual questions are as follows:

What is your regular hourly rate of pay on this job?

During the last year, how many times did you see or talk to a medical doctor?

Do you have an internet connection in your home?

The question must exactly define the fact it intends to measure. Experience has shown that even a small change in the question text may lead to a substantially different answer. For example, take the question about the number of doctor visits: "During the last year, how many times did you see or talk to a medical doctor?" Does this also include visits for other than medical reasons? And what about visits with ill children?

*Nonfactual questions* ask about attitudes and opinions. An *opinion* usually reflects a view on a specific topic, like voting intention for the next election. An *attitude* is a more general concept, reflecting views about a wider, often more complex issue. With opinions and attitudes, there is no such thing as a true value. They measure a subjective state of the respondent that cannot be observed by other means. The opinion or attitude only exists in the mind of the respondent.

To avoid surprises, it is good to realize how opinions come into being. There are various theories explaining how respondents determine their answers to opinion questions. One such theory is the *online processing model* described by Lodge et al. (1995). According to this theory, people maintain an overall impression of ideas, events, and persons. Every time they are confronted with new information, and their summary view is updated spontaneously. When they have to answer an opinion question, their response is determined by this overall impression. The online processing model should typically be applicable to opinions about politicians and political parties.

There are situations in which people have not yet formed an opinion about a specific issue. They only start to think about it when confronted with the question. According to the *memory-based model* of Zaller (1992), people collect all kinds of information from the media and in contacts with other people. Much of this information is stored in memory without paying attention to it. When respondents have to answer an opinion question, they may recall some of the relevant information stored in memory. Due to the limitations of the human memory, only part of the information is used. This is the information that immediately comes to mind when the question is asked. This is often the information that only recently has been stored in memory. Therefore, the memory-based model can explain why people seem to be unstable in their opinions. Their answer will often be determined by the way the issue was recently covered in the media.

## 3.3 THE TEXT OF THE QUESTION

The question text is probably the most important part of the question. This is what the respondents respond to. If they do not understand the question, they will not give the correct answer, or they will give no answer at all. The following set of checks may help us to decide whether a question text has defects that may give rise to a wrong answer.

### 3.3.1 Is Familiar Wording Used in the Text?

The question text must use words that are familiar to those who have to answer the question. Particularly, questionnaire designers must be careful not to use jargon that is familiar to themselves, but not to the respondents. Economists may understand a question like

> Do you think that food prices are increasing at the same rate as a year ago, at a faster rate, or at a slower rate?

This question asks about the rate at which prices rise, but a less-knowledgeable person may easily interpret the question as asking whether prices decreased, stayed the same, or increased.

Unnecessary and possibly unfamiliar abbreviations must be avoided. Do not expect respondents to be able to answer questions about, for example, caloric content of food, disk capacity (in gigabytes) of their computer, or the bandwidth (in Mbps) of their internet connection.

Indefinite words (sometimes also called *vague quantifiers*) like *usually*, *regularly*, *frequently*, *often*, *recently*, and *rarely* must be avoided if there is no additional text explaining what they exactly mean. How regular is regularly? How frequent is frequently? These words do not have the same meaning for every respondent. One respondent may interpret *regularly* as every day, whereas it could mean once a month to another respondent. Here is an example of a question with an indefinite word:

Have you been to the cinema recently?

What does *recently* mean? It could mean the last week or the last month. A better question specifies the time period:

In the last week, have you been to the cinema?

Even this question text could cause some confusion. Does *last week* mean the last seven days or maybe the period since the last Sunday?

### 3.3.2 Is the Question Ambiguous?

If the question text is interpreted differently by different respondents, their answers will not be comparable. For example, if a question asks about income, it must be clear whether it is about weekly, monthly, or annual income. Moreover, it must be clear whether respondents must specify their income before or after tax has been deducted.

Vague wording may also lead to interpretation problems. Suppose, respondents are confronted with the following question:

Are you satisfied with the recreational facilities in your neighborhood?

They may wonder what recreational facilities exactly are. Is this a question about parks and swimming pools? Do recreational facilities also include libraries, theatres, cinemas, playgrounds, dance studios, and community

Which of the following statements is closest to your view?

○  I have a lot of sympathy with young Muslims who leave the UK the join fighters in Syria.
○  I have some sympathy with young Muslims who leave the UK the join fighters in Syria.
○  I have no sympathy with young Muslims who leave the UK the join fighters in Syria.
○  Do not know

FIGURE 3.1   An ambiguous question.

centers? What will respondents have in their mind when they answer this question? It is better to explain exactly in the question text what recreational facilities are.

A poll in the English newspaper *The Sun* contained another example of an ambiguous question. This poll was conducted in November 2015 among British Muslims. One of the questions in the poll is shown in Figure 3.1.

From the answers to this question, the newspaper concluded that 1 in 5 British Muslims have sympathy for jihadis. There are, however, at least two problems with this question. The first one is with the word *sympathy*. What does it mean when one has sympathy? Does it mean that they are doing the right thing? Or can it mean that one understands their feelings and motives but does not agree with what they are doing? Another problem with this question is that it was about fighters in Syria in general, and not just about jihadi fighters. There were several groups of fighters, including British exservicemen fighting against ISIS with the Kurds, and anti-Assad Muslim forces who are also fighting against ISIS.

### 3.3.3  Is the Question Text Too Long?

The question text should be as short as possible. It is not easy to comprehend a long text. It is not unlikely that part of the text is ignored, which may change the meaning of the question. Long texts may also cause *respondent fatigue*, resulting in a decreased motivation to continue. Of course, the question text should not be so short that it becomes ambiguous. Here is an example of a question that may be too long:

> During the past seven days, were you employed for wages or other remuneration, or were you self-employed in a household enterprise, were you engaged in both types of activities simultaneously, or were you engaged in neither activity?

Some indication of the length and difficulty of a question text can be obtained by counting the total number of syllables and the average number

TABLE 3.2   Indicators for the Length and Complexity of a Question

| Questions | Words | Syllables | Syllables per Word |
|---|---|---|---|
| Have you been to the cinema in the last week? | 9 | 12 | 1.3 |
| Are you satisfied with the recreational facilities in your neighborhood? | 10 | 21 | 2.1 |
| During the past seven days, were you employed for wages or other remuneration, or were you self-employed in a household enterprise, were you engaged in both types of activities simultaneously, or were you engaged in neither activity? | 38 | 66 | 1.7 |

of syllables per word. Table 3.2 gives examples of these indicators for three question texts. The first question is simple and short. The second one is also short, but it is more difficult. The third question is very long and has an intermediate complexity.

It should be noted that research shows that longer question texts sometimes also lead to better answers. If an interviewer in a face-to-face poll or a telephone poll reads out a long text, the respondent has more time to think about the correct answer. Moreover, according to Kalton and Schuman (1982), a longer text may work better for open questions about threatening topics.

### 3.3.4  Is It a Recall Question?

Recall questions are a source of problems. It is not easy to answer a question requiring recall of events that have happened a long time ago (a *recall question*). The reason is that people make *memory errors*. They tend to forget events, particularly when they happened a long time ago. Recall errors are more severe as the length of the reference period is longer. Important events, more interesting events, and more frequently happening events will be remembered better than other events. Here is an example of a recall question:

> In the last two years, how many times did you contact your family doctor?

This is a simple question to ask, but for many people difficult to answer. Recall errors may even occur for shorter time periods. In the 1981 Health Survey of Statistics Netherlands, respondents had to report contacts with their family doctor over the last three months. Sikkel (1983) investigated the memory effects. It turned out that the percentage of unreported contacts increased with time. The longer ago an event took place, the more likely it was forgotten. The percentage of unreported events for this question increased on average with almost 4% per week. Over the total period of three months, about one quarter of the contacts with the family doctor were not reported.

Recall questions may also suffer from *telescoping*. This occurs if events are reported as having occurred either earlier or later than they actually did. As a result, an event is incorrectly included within the reference period, or incorrectly excluded from the reference period. Bradburn et al. (2004) note that telescoping leads more often to overstating than to understating the number of events. Particularly, for short reference periods, telescoping may lead to substantial errors in estimates.

### 3.3.5 Is It a Leading Question?

A *leading question* is a question that is not asked in a neutral way, but that leads the respondents in the direction of a specific answer. It will be clear that leading questions are a sign of a bad poll. Here is an example of such a question:

> Do you agree with the majority of people that the quality of the health care in the country is falling?

This question contains a reference to the *majority of people*. It suggests it is socially undesirable to disagree. A question can also become leading by including the opinion of experts in questions text, as is in the following question:

> Most doctors say that cigarette smoking causes lung cancer. Do you agree?

The question text should not contain loaded words that have a tendency of being attached to extreme situations. Here is an example:

> What should be done about murderous terrorists who threaten the freedom of good citizens and the safety of our children?

Particularly, adjectives like *murderous* and *good* increase a specific loading of the question.

TABLE 3.3    Offering More Information in the Question Text

| Question 1 | Question 2 |
|---|---|
| An increase in the powers of the European Parliament will be at the expense of the national Parliament. Do you think the powers of the European Parliament should be increased? | Many problems cross national borders. For example, 50% of the acid rain in The Netherlands comes from other countries. Do you think the powers of the European Parliament should be increased? |

Opinion questions can address topics about which respondents have not yet made up their mind. They could even lack sufficient information for a balanced judgment. To help them, a question can give additional information in the question text. Such information should be objective and neutral and should not lead respondents in a specific direction. Saris (1997) performed an experiment showing the dangers of adding information to the question text. He measured the opinion of the Dutch about increasing the power of the European Parliament. Respondents were randomly assigned one of the two questions in Table 3.3.

For the question on the left, 33% answered *yes*, and 42% answered *no*. And for the question on the right, 53% answered *yes*, and only 23% answered *no*. So the percentages differed substantially. It is not surprising that the percentage *yes* is much lower on the left than on the right, because the question on the left stresses a negative aspect of the European Parliament, and the question on the right a positive point.

### 3.3.6 Does the Question Ask Things People Don't Know?

A question text can be very simple, and completely unambiguous, but it can still be impossible to answer it. This may happen if respondents are asked for facts they do not know. Such questions should not appear in the questionnaire. Here is an example:

> How many hours did you listen to your local radio station in the last six months?

People do not keep record of all kinds of simple things happening in their life. So, they can only make a guess. This guess need not necessarily be an accurate one. Answering such a question is even more difficult if it is about a relatively long reference period.

### 3.3.7 Is It a Sensitive Question?

Be careful with the answers to sensitive questions. *Sensitive questions* address topics that respondents may see as embarrassing. This can result in inaccurate answers. They will avoid socially undesirable answers and give socially more acceptable answers, or they give no answer at all. Examples of sensitive questions are questions about topics such as income, health, criminal behavior, or sexual behavior. A researcher should avoid sensitive questions as much as possible. Of course, this cannot be avoided if the poll is about a sensitive topic.

Sensitive questions can be asked in such a way that the likelihood of a correct response is increased, and a more honest response is facilitated. One approach is including the question in a series of less-sensitive questions about the same topic. Another approach is making it clear in the question text that a specific behavior or attitude is not so unusual. Bradburn et al. (2004) give the following example:

> Even the calmest parents sometimes get angry at their children. Did your children do anything in the past seven days to make you angry?

A question can also be made less sensitive by referring in the question text to experts who find the behavior not so unusual:

> Many doctors now believe that moderate drinking of liquor helps us to reduce the likelihood of heart attacks and strokes. Did you drink any liquor in the past month?

One should, however, be careful with making questions less sensitive. There is always a risk that such modifications result in leading questions.

A question asking for numerical quantities (like income) can be considered threatening if an exact value must be supplied. This can be avoided by letting respondents select a range of values. For example, instead of asking for someone's exact income, he or she can be asked to choose an income category.

### 3.3.8 Is It a Double Question (Also Called a Double-Barreled Question)?

A question must ask one thing at the time. If more than one thing is asked in a question, and only one answer can be given, it is unclear what the answer means. Here is an example of such a double question:

> Do you think that people should eat less and exercise more?

It actually consists of the following two questions:

Do you think that people should eat less?

Do you think that people should exercise more?

Suppose, a respondent thinks that people should eat less, but not exercise more. What would his answer be? "Yes" or "no?" The solution to this problem is simple: the question must be split into two questions each asking one thing at the time. The questionnaire should not contain double questions.

### 3.3.9 Is It a Negative Question?

Questions must not be asked in the negative as this is more difficult to understand. For example, people may be confused by this question:

Are you against a ban on smoking?

Even more problematic are double-negative questions. They are a source of serious interpretation problems. Here is an example:

Would you rather not use a nonmedicated shampoo?

Negative questions can usually be rephrased such that the negative effect is removed. For example, "are you against a ban ..." can be replaced by "are you in favor ...." The conclusion is clear: negative, and certainly double-negative questions should be avoided.

### 3.3.10 Is It a Hypothetical Question?

It is difficult to answer questions about imaginary situations, as they relate to circumstances people have never experienced. At best, their answer is a guesswork, and a total lie at worst. Following is an example of a hypothetical question:

If you were the president of the country, how would you stop crime?

Hypothetical questions are often asked to get more insight in attitudes and opinions of people about certain issues. However, little is known about processes in the mind that lead to an answer to such questions. So, one may wonder whether hypothetical questions really measure what a researcher wants to know. It may be better to avoid such questions.

## 3.4 QUESTION TYPES

Respondents filling in the questionnaire will be confronted with different types of questions. They will, for example, see open questions, closed questions (with one possible answer), closed questions for which more than one answer is allowed, and numerical questions. Every question type has its advantages and disadvantages. A bad question design can even lead to wrong answers. To assess the quality of the questionnaire, it is therefore a good idea to take a critical look to the format of the questions.

### 3.4.1 Open Questions

An *open question* is a simple question to ask. Respondents can answer such a question completely in their own words. An open question is typically used in situations where people should be able to express themselves freely. Open questions invoke spontaneous answers. But open questions also have disadvantages. The possibility always exists that respondents overlook an answer possibility. Consider the question in Figure 3.2 which was taken from a readership poll.

Research in The Netherlands has shown that if this question is offered as an open question, typically television guides are overlooked. If a list is presented containing all weekly's, including television guides, much more people report having read TV guides.

Asking an open question may also lead to vague answers. Figure 3.3 shows an example. It will be unclear for many respondents what kind of answer is expected. They could answer something like *salary* or *pay*. And what do they mean when they say this? Is it important to get a high salary, or a regular salary, or maybe both?

---

In the last two weeks, which weekly magazines have you read.
.........................................................................................................................................................

---

FIGURE 3.2   An open question.

---

What do you consider the most important aspect of your job?
.........................................................................................................................................................

---

FIGURE 3.3   A vague open question.

The open questions in Figures 3.2 and 3.3 allow for an answer consisting of at most one line of text. One could also offer a larger box with the possibility of more lines of text. The researcher should realize that the size of the answer box determines the length of the answer. A large box will produce longer answers, and a small box only short answers. So, the size of the box should reflect what the researcher wants to have, a short note, or an extensive description.

Processing answers to open questions is cumbersome, particularly if the answers are written down on a paper form. Entering such answers in the computer takes effort, and even more if the written text is not very well readable. Furthermore, analyzing answers to open questions is not very straightforward. It often has to be done manually because there is no intelligent software that can do this automatically.

To avoid the above-mentioned potential problems, questionnaires should contain as few open questions as possible. However, there are situations in which there is no alternative for an open question. An example is a question asking for the respondent's occupation. A list containing all possible occupations would be very long. It could easily have thousands of entries. All this makes it very difficult, if not impossible, to locate a specific occupation in a list. The only way out is to ask for occupation by means of an open question. Then, the answers have to be processed, and this calls for extensive, time-consuming automatic and/or manual coding procedures. There are more advanced coding techniques, like hierarchical coding and trigram-coding. See, for example, Roessingh and Bethlehem (1993).

### 3.4.2 Closed Question, One Answer

A *closed question* measures a *qualitative variable*. It is a variable that divides people into groups. The variable gender divides people into males and females, and the variable voting intention (in the United States) could divide people into the groups Democrats, Republicans, Other parties, and Nonvoters.

There is always a list of answer options for a closed question. These options must correspond to the groups the qualitative variable distinguishes. It is important that the answer options are exhaustive and nonoverlapping. Each respondent must be able to find the one option that fits his situation. Figure 3.4 contains an example of a closed question.

As respondents can only be a member of one group, they must select only one answer option in the list. For paper questionnaires, there is a risk

| What is your present marital status? |
| :--- |
| ○ Never married<br>◉ Married<br>○ Separated<br>○ Divorced<br>○ Widowed |

FIGURE 3.4    A closed question.

that someone checks more than one option. Then, it will be difficult to determine the right one. For online questionnaires, such a problem does not exist. Online questionnaires use radio buttons for closed questions. Only one radio button can be selected. Choosing another option automatically deselects the currently selected option.

There will be problem if a respondent cannot find the proper answer. One way to avoid this problem is to add the category *Other*. It may be useful to give respondents the possibility to explain which other situation applies to them. To that end, there could be a text field linked to *Other*. An example is the question in Figure 3.5.

If the list with answer options is long, it will be difficult to find the proper answer. The closed question in Figure 3.6 is an example of such a question with many options.

Long option lists may cause problems in face-to-face and telephone polls. For these polls, the interviewer reads out all options when asking a closed question. By the time the interviewer reaches the end of the list, the respondent may have already forgotten the first options in the list. This causes a preference for an answer near the end of the list. This is called a *recency effect*.

Use of *show cards* may help in face-to-face interviews. Such a card contains the complete list of possible answers to a question. The interviewer hands over the card to the respondents, and they can pick the proper answer from the list.

| To which type of program do you listen to the most on your local radio station? |
| :--- |
| ○ Music<br>◉ News and current affairs<br>○ Sports<br>○ Culture<br>○ Other, please specify: ............................................................................................ |

FIGURE 3.5    A closed question with an option *other*.

Which means of transport do you use most for traveling within your town?

○ Walking
◉ Bicycle
○ Electric bicycle
○ Moped
○ Scooter
○ Motorcycle
○ Car
○ Bus
○ Tram
○ Metro
○ Light-rail
○ Train
○ Other means of transport

FIGURE 3.6    A closed question with many options.

In the case of an online poll or a mail poll, respondents have to read the list themselves. This is not always done very carefully because they may quickly lose interest. This causes a tendency to pick an answer early in the list. This is called a *primacy effect*.

A special kind of closed question is the *rating scale question*. Such a question measures the opinion of attitude of a person with respect to a certain issue. Instead of offering a choice between two options (e.g., *Agree* and *Disagree*), a so-called *Likert scale* can be used. This is a set of options giving people the possibility to express the strength of their opinion or attitude. The Likert scale was invented by Likert Rensis in 1932. A Likert scale question often has five categories. Figure 3.7 contains an example.

Rating scale questions can have a different number of options, for example, three options or seven options. Three options may restrict people too much in expressing their feelings, and seven or more options may be too much. Then, it will be hard to find the option that is closest to one's feelings.

Rating scale questions should preferably have an odd number of options. There should be one neutral option in the middle. This option is for those without a clear opinion about the particular issue. A drawback of

Taking all things together, how satisfied or dissatisfied are you with life in general?

○ Very dissatisfied
○ Dissatisfied
◉ Neither dissatisfied, nor satisfied
○ Satisfied
○ Very satisfied

FIGURE 3.7    A rating scale question.

> Do you remember for sure whether or not you voted in the last elections for the European Parliament of May 22–24, 2014?
>
> ○ Yes, I voted
> ○ No, I did not vote
> ◉ Do not know

FIGURE 3.8   A closed question with *do not know*.

this middle option is that respondents may pick this option as a way-out, as a way to avoid having to express an opinion. It can be seen as a form of *satisficing*. This term was introduced by Krosnick (1991). It is the phenomenon that respondents do not do all they can to provide the correct answer. Instead, they give a more or less acceptable answer with minimal effort.

Sometimes, people cannot answer a question because they simply do not know the answer. They should have the possibility to indicate this on the questionnaire form, as forcing them to make up an answer will reduce the validity of the data. It has always been a matter of debate how to deal with the *don't know* option. One way to deal with *don't know* is to offer it as one of the options in a closed question. See Figure 3.8 for an example.

Particularly in mail and online polls, such a question may suffer from satisficing. People seek the easiest way to answer the question by simply selecting the *don't know* option. Such behavior can be avoided in face-to-face and telephone polls. Interviewers can be trained in assisting respondents to give a *real* answer and to avoid *don't know* as much as possible. The option is not explicitly offered but is implicitly available. If respondents persist they really do not know the answer, the interviewer records this response as *don't know*.

Another way to avoid satisficing is by including a *filter question*. This question asks the respondents whether they have an opinion on a certain issue. Only if they say they have, they are asked to specify their opinion in the subsequent question.

A final example of a closed question with one answer shows the dangers of not properly implementing such a question (see Figure 3.9). It is a question from the Standard Eurobarometer 44.1 (1995). The Eurobarometer is a poll that is conducted regularly in the member states of the European Union. The European Commission is responsible for the Eurobarometer. This question, about a possible extension of the European Union with some Central and East European countries, was asked in 1995.

The Eurobarometer is a face-to-face survey. The interviewers read out the question text and the answer options. For this question, they were

Some say the countries of Central and Eastern Europe, such as the Czech Republic, Hungary, Poland, and Slovakia, should become member states of the European Union. What is your opinion on this ? Should they become members?

○ in less than 5 years
○ in the next 5 to 10 years
○ in over 10 years

○ I do not think these countries should become members of the European Union (Spontaneous)
○ Do not know

FIGURE 3.9   A closed question with unbalanced answer options.

instructed to read out only the first three answer options. So, one could only agree with the extension of the European Union with the countries mentioned. Only if someone spontaneously insisted these countries should not become a member of the European Union, the interviewer could record this in the questionnaire. As a result of this approach, the answers were biased in the direction of extension of the European Union.

One could also have mixed feelings about the text of the question. It starts with the sentence "Some say the countries of Central and Eastern Europe, such as the Czech Republic, Hungary, Poland, and Slovakia, should become member states of the European Union." This suggests that it is socially desirable to be in favor of extension of the European Union.

### 3.4.3  Closed Question, More Than One Answer

The closed question described so far allows people to select exactly one answer option. All answer options are supposed to be mutually exclusive and exhaustive. Therefore, people can always find the one option describing their situation. Sometimes, however, there are closed questions in which respondents are given the possibility to select more than one option, simply because more options apply to them. Such a question is called a *check-all-that-apply question*. Figure 3.10 shows an example.

What are your normal modes of transport to work?
*Check all that apply*

☑ Walking
☑ Bicycle
☐ Motorcycle
☑ Car
☐ Bus, tram
☐ Other mode of transport

FIGURE 3.10   A closed question with more than one answer.

| What are your normal modes of transport to work? |
|---|
| Yes  No |
| ⊙  ○  Walking |
| ⊙  ○  Bicycle |
| ○  ⊙  Motorcycle |
| ⊙  ○  Car |
| ○  ⊙  Bus, tram |
| ○  ⊙  Other mode of transport |

FIGURE 3.11    A closed question in an alternative format.

This question asks for the respondent's mode of transport to work. It is not unlikely that he uses more than one mode of transport for his journey: First, he goes to the railway station by bicycle, then he takes the train into town, and finally he walks from the railway station to the office. It is clear that the question must be able to record all these answers. Therefore, it must be possible to check every option that applies.

Check-all-that-apply questions in online polls use square *check boxes* to indicate that several answers can be selected. Dillman et al. (1998) have shown that such a question may lead to problems if the list of options is very long. Respondents may stop after having checked a few answers and do not look at the rest of the list any more. So, too few options are checked. This can be avoided by using a different format for the question. An example is presented in Figure 3.11.

Each check box has been replaced by two radio buttons, one for *Yes* and the other for *No*. This approach forces respondents to do something for each option. They either have to check *Yes* or *No*. So, they have to go down the list option by option and give an explicit answer for each option. This approach causes more options to be checked, but it has the disadvantage that it takes more time to answer the question.

### 3.4.4 Numerical Question

Another frequently occurring type of question is a *numerical question*. The answer to such a question is simply a number. Examples are questions about age, income, or prices. Moreover, in many household polls, there is a question about the number of members in the household. See Figure 3.12 for an example.

| How many people are there in your household? |
|---|
| _ _ people |

FIGURE 3.12    A numerical question.

> In the last seven days, how many hours did you listen to the radio?
>
> _ _ _ hours

FIGURE 3.13    A hard to answer numerical question.

> In the last seven days, how many hours did you listen to your local radio station?
>
> ○  Less than 1 hour
> ○  At least 1 hour, but less than 2 hours
> ○  At least 2 hours, but less than 5 hours
> ○  At least 5 hours, but less than 10 hours
> ◉  10 hours or more

FIGURE 3.14    A numerical question transformed into a closed question.

The two separate dashes give a visual clue as to how many digits are (at most) expected. Numerical questions in electronic questionnaires may have a lower and upper bound built in for the answer. This ensures that entered numbers are always within the valid range.

Note that in many situations it is not possible to give an exact answer to a numerical question, because respondents simply do not know the answer. An example is the question in Figure 3.13 in which they are asked to specify how many hours they listened to the radio in the last seven days.

An alternative may be to replace the numerical question by a closed question, in which the answers options correspond to intervals. See Figure 3.14 for an adapted version of the question in Figure 3.13.

This question has five possible answers. Of course, different classifications are possible. Such a classification can be more detailed (more options), or less detailed (less options). If there are many options, it may be difficult for respondents to select the true one. If there are only a few options, the answers are less informative. Note that a question with a detailed classification may also suffer from satisficing. To avoid having to find the true category, respondents just select a category in the middle of the classification.

A special type of numerical question is a *date question*. Many polls ask respondents to specify dates, for example, date of birth, date of purchase of a product, or date of retirement. Figure 3.15 contains an example of a date question.

| What is your date of birth? |
| :--- |
| $--$   $--$   $----$ |
| *day*   *month*   *year* |

FIGURE 3.15   A date question.

Of course, a date could also be asked by means of an open question, but if used in interview software, dedicated date questions offer much more control, and thus less errors will be made in entering a date.

### 3.4.5 Grid Question

If there is a series of questions with the same set of possible answers, they can be combined into a *grid question.* Such a question is sometimes also called a *matrix question.* A grid question is a kind of table in which each row represents a single question and each column corresponds to one possible answer option. Figure 3.16 shows a simple example of a grid question. In practice, grid questions are often much larger than this one.

At first sight, grid questions seem to have some advantages. A grid question takes less space in the questionnaire than a set of single questions. Moreover, it provides respondents with more oversight. Therefore, it can reduce the time it takes to answer the questions. Couper et al. (2001) indeed found that a gird question takes less time to answer than a set of single questions.

According to Dillman et al. (2009), answering a grid question is a complex cognitive task. It is not always easy for respondents to link a single question in a row to the proper answer in a column. Moreover, they can navigate through the grid in several ways: row-wise, column-wise, or a mix. This increases the risk of missing answers to questions, resulting in unanswered questions (also called item nonresponse).

|  | Excellent | Very good | Good | Fair | Poor |
| :--- | :---: | :---: | :---: | :---: | :---: |
| How would you rate the overall quality of the radio station? | ○ | ○ | ◉ | ○ | ○ |
| How would you rate the quality of the news programs? | ○ | ◉ | ○ | ○ | ○ |
| How would you rate the quality of the sport programs? | ○ | ◉ | ○ | ○ | ○ |
| How would you rate the quality of the music programs? | ○ | ○ | ○ | ◉ | ○ |

FIGURE 3.16   A grid question.

| | Excellent | Very good | Good | Fair | Poor |
|---|:---:|:---:|:---:|:---:|:---:|
| How would you rate the overall quality of the radio station? | ○ | ○ | ◉ | ○ | ○ |
| How would you rate the quality of the news programs? | ○ | ○ | ◉ | ○ | ○ |
| How would you rate the quality of the sport programs? | ○ | ○ | ◉ | ○ | ○ |
| How would you rate the quality of the music programs? | ○ | ○ | ◉ | ○ | ○ |

FIGURE 3.17   A grid question with straight-lining.

Dillman et al. (2009) advise to limit the use of grid questions in online polls as much as possible. And if they are used, the grid should neither be too wide nor be too long. Preferably, the whole grid should fit on a single screen (in the case of on online questionnaire). This is not so easy to realize as different respondents may have set different screen resolutions on their computer screens or use different devices (laptop, tablet, and smartphone). If they have to scroll, either horizontally or vertically, they may easily get confused, leading to wrong or missed answers.

Some experts, for example Krosnick (1991) and Tourangeau et al. (2004), express concern about a phenomenon that is sometimes called *straight-lining*. This particularly occurs for grid questions in online polls. Respondents take the easy way out by giving the same answer to all questions in the grid. They simply check all radio buttons in the same column. Often this is the column corresponding to the middle (neutral) response option. Figure 3.17 contains an example of straight-lining.

Straight-lining the middle answer option can be seen as a form of satisficing. It is a means of quickly answering a series of questions without thinking. It manifests itself in very short response times. So, short response times for grid questions (when compared to a series of single questions) are not always a positive message. It can mean that there are measurement errors caused by satisficing.

If a grid question is used, it must have a good visual layout. For example, a type of shading as in the above-mentioned examples reduces confusion and therefore also reduces item nonresponse. *Dynamic shading* may help for a grid question in an online poll. Kaczmirek (2010) distinguishes preselection shading and postselection shading. *Preselection shading* comes down to changing the background color of a cell or row of the matrix question if the cursor is moved over it by the respondent. Preselection shading helps respondents to locate the proper answer to a question. It is active before the answer is clicked. *Postselection shading* means shading of a cell or row in the grid after the answer has been selected. This feedback informs the respondent which answer to which question was selected. Kaczmirek (2010) concluded that preselection and postselection shading of complete rows improves the quality of the answers. However, preselection shading of just the cell reduced the quality of the answers.

Galesic et al. (2007) also experimented with postselection shading. The font color or the background color was changed immediately after respondents answered a question in the matrix. This helped respondents to navigate and therefore improved the quality of the data.

## 3.5 THE ORDER OF THE QUESTIONS

Questions appear in a questionnaire in a specific order. This question order may affect the way questions are answered. Therefore, it is not only important to look at each question separately, but also to the context in which the questions are asked. This section discusses some aspects of question order.

A first aspect of question order is *grouping* of questions. Usually, it is a good idea to keep questions about the same topic close together. This will make it easier to answer questions. Therefore, this will improve the quality of the answers.

A second aspect is a potential *learning effect* or *framing*. An issue addressed early in the questionnaire may make people think about it. This may affect answers to later questions. As an example, this phenomenon played a role in a Dutch housing demand survey. People turned out to be much more satisfied with their housing conditions if this question was asked early in the questionnaire. The questionnaire contained a number of questions about the presence of all kinds of facilities in and around the house (Do you have a bath? Do you have a garden? Do you have a central heating system?). As a consequence, several people realized that they lacked these facilities and therefore became less and less satisfied with their housing conditions in the course of the interview.

Question order can affect results in two ways. One is that mentioning something (an idea, an issue, or a brand) in one question can make people think of it while they answer a later question, when they might not have thought of it if it had not been previously mentioned. In some cases, this problem may be reduced by randomizing the order of related questions. Separating related questions from unrelated ones might also reduce this problem, though neither technique will completely eliminate it.

Tiemeijer (2008) describes an example in which the answers to a specific question were affected by a previous question. The Eurobarometer is an opinion poll that is regularly conducted in all member states of the European Union since 1973. The following question was asked in 2007:

> Taking everything into consideration, would you say that the country has on balance benefited or not from being a member of the European Union?

It turned out that 69% of the respondents said that their country had benefited from the European Union. A similar question was included at the same time in a Dutch opinion poll (*Peil.nl*). However, the question was preceded by another question that asked respondents to select the most important disadvantages of being a member of the European Union. Among the items in the list were the fast extension of the EU, the possibility of Turkey becoming a member state, the introduction of the Euro, the waste of money by the European Commission, the loss of identity of the member states, the lack of democratic rights of citizens, veto rights of member states, and possible interference of the European Commission with national issues. The result was that only 43% of the respondents considered membership of the European Union beneficial. Apparently, the question about disadvantages increased negative feelings about the European Union.

A third aspect of question order is that a specific question order can encourage people to complete the questionnaire. Ideally, the early questions in a poll should be easy and pleasant to answer. Such questions encourage people to continue the poll. Whenever possible, difficult or sensitive questions should be asked near the end of the questionnaire. If these questions cause people to quit, at least many other questions have been answered.

Another aspect of question order is *routing*. Usually, not every question is relevant for each respondent. For example, an election poll may contain questions for both voters and nonvoters. For voters, there may be questions about party preference, and for nonvoters there may be questions about the reasons for not voting. Irrelevant questions may cause irritation, possibly resulting in refusal to continue. Moreover, they may not be able to answer questions not relating to their situation. Finally, it takes more time to complete a questionnaire if also irrelevant questions must be answered. To avoid all these problems, there should be *routing instructions* in the questionnaire. Figure 3.18 contains an example of a simple questionnaire with routing instructions.

**1. Generally speaking, how much interest would you say you have in politics?**
- O A great deal
- O A fair amount
- O Only a little
- O No interest at all

**2. What is your age (in years)?** __ __ __

*Answer questions below only if you are 18 years or older.*

**3. Of the issues that were discussed during the election campaign, which one was most important for you?**

......................................................................................................................................................................

**4. Did you vote in the last parliamentary elections?**
- O Yes     *Go to question 6*
- O No      *Go to question 5*

**5. Why did you not vote?**
- O Not interested
- O Had no time
- O Forgot about it
- O Too young
- O Other reason
*Go to question 9*

**6. Which party did you vote for?**
- O Christian-Democrat Party
- O Social-Democrat Party
- O Liberal Party
- O Green Party
- O Other party

**7. Which other party did you consider (*check all that apply*)?**
- ☐ Christian-Democrat Party
- ☐ Social-Democrat Party
- ☐ Liberal Party
- ☐ Green Party
- ☐ Other party

**8. Do you think that voting for parliamentary elections should be mandatory, or do you think that people should only vote if they want to?**
- O Strongly favor compulsory voting
- O Favor compulsory voting
- O Favor people voting only if they want to
- O Strongly favor people voting only if they want to

**9. Did you follow the election campaign in the media (*check all that apply*)?**
- ☐ Yes, on radio
- ☐ Yes, on TV
- ☐ Yes, on internet

*End of questionnaire*

FIGURE 3.18   A simple election poll with routing instructions.

There are two types of routing instructions. The first type is that of a jump instruction attached to an answer option of a closed question. Question 4 has some instructions: if a respondent answers *Yes*, he is instructed to jump to question 6, and continue from there. If his answer to question 4 is *No*, he is instructed to go to question 5.

Sometimes, a routing instruction does not depend on just an answer to a closed question. It may happen that the decision to jump to another question depends on the answer to several other questions, or on the answer to another type of question. In this case, a route instruction may take the form of an instruction between the questions. This is a text placed between questions. Figure 3.16 contains such an instruction between questions 2 and 3. If a respondent is younger than 18, the rest of the questionnaire is skipped.

We already mentioned that routing instructions not only see to it that only relevant questions are asked, but they also reduce the number of questions asked, so that the interview takes less time. However, it should be noted that many- and complex-route instructions increase the burden for the interviewers. This complexity may be an extra source of errors.

## 3.6  TESTING THE QUESTIONNAIRE

Before a questionnaire can be used for collecting data, it must be tested. Errors in the questionnaire may cause wrong questions to be asked, and right questions to be skipped. Moreover, errors in the questions themselves may lead to errors in answers. Every researcher will agree that testing is important, but this does not always happen in practice. Often there is no time to carry out a proper test. An overview of some aspects of questionnaire testing is given here. More information can be found, for example, in Converse and Presser (1986).

Questionnaire testing usually comes down to trying it out in practice. There are two approaches to do this. One is to imitate a normal interview situation. Interviewers make contact with respondents and interview them, like in a real interview situation for a poll. The respondents do not know it is just a test, and therefore they behave like what they would do during a normal interview. If they know it was just a test, they could very well behave differently.

Another way to test a questionnaire is to inform respondents that they are part of a test. This has the advantage that the interviewers can ask the respondents whether they have understood the questions, whether things were unclear to them, and why they gave specific answers. This is sometimes called *cognitive interviewing.*

A number of aspects of a questionnaire should be tested. Maybe the most important aspect is the validity of the question. Does the question really measure what the researcher wants it to measure? It is not simple to establish question validity in practice. The first step may be to determine the meaning of the question. It is important that the researcher and the respondent interpret the question exactly in the same way. There are ample examples in the questionnaire design literature about small and large misunderstandings. For example, Converse and Presser (1986) mention a question about *heavy traffic in the neighborhood*, in which the researcher meant trucks and respondents thought the question was about drugs. Another question asked about *family planning*, in which the researcher meant birth control and respondents did interpret this as saving money for vacations.

The above-mentioned examples make clear how important validity testing is. Research has shown that often respondents interpret questions differently than the researcher intended. Moreover, if respondents do not understand the question, they change the meaning of the question in such a way that it can be answered by them.

Another aspect of questionnaire testing is to check whether questions offer sufficient variation in answer possibilities. A poll question is not very interesting for analysis purposes if all respondents give the same answer. It must be possible to explore and compare the distribution of the answers to a question for several subgroups of the population.

It should be noted that there are situations in which a very skew answer distribution may be interesting. For example, De Fuentes-Merillas et al. (1998) investigate addiction to scratch cards in The Netherlands. It turned out that only 0.24% of the adult population was addicted. Although this was a very small percentage, it was important to have more information about the size of the group.

The meaning of a question may be clear, and it also may allow sufficient variation in answers, but this still not means that it can always be answered. Some questions are easy to ask, but difficult to answer. A question like "In your household, how many kilograms of coffee did you consume in the last year?" is clear, but very hard to answer because respondents simply do not know the answer, or can determine the answer only with great effort. Likewise, asking for the net yearly income is not as simple as it looks. Researchers should realize that they may get only an approximate answer.

Many people are reluctant to participate in polls. Moreover even if they corporate, they may not be very enthusiastic or motivated to answer the questions. Researchers should realize that this may have an effect on the quality of the answers given. The more interested respondents are, the better their answers will be. One aspect of questionnaire testing is to determine how interesting questions are for respondents. The number of uninteresting questions should be as small as possible.

Another important aspect is the length of the questionnaire. The longer the questionnaire, the larger the risk of problems. *Questionnaire fatigue* may cause respondents to stop answering questions before they reach the end of the questionnaire. A rule sometimes suggested in The Netherlands is that an interview should not last longer than a class in school (50 minutes). However, it should be noted that this also partly depends on the mode of interviewing. For example, telephone interviews should take less time than face-to-face interviews. And completing a questionnaire online should take not more than 15 minutes.

Not only the individual questions, but also the structure of the questionnaire as a whole, must be tested. Each respondent follows a specific route through the questionnaire. The topics encountered en route must have a meaningful order for all respondents. One way the researcher can check this is to read out loud the questions (instead of reading them). While listening to this story, unnatural turns will become apparent.

To keep the respondent interested, and to avoid questionnaire fatigue, it is recommended to start the questionnaire with interesting questions. Uninteresting and sensitive questions (gender, age, and income) should come at the end of the questionnaire. In this way, potential problems can be postponed until the end.

It should be noted that sometimes the structure of the questionnaire requires uninteresting questions like gender to be asked early in the questionnaire. This may happen if they are used as filter questions. The answer to such a question determines the route through the rest of the questionnaire. For example, if a questionnaire contains separate parts for male and female respondents, first gender of the respondent must be determined.

CAPI, CATI, and online polls allow for various forms of computer-assisted interviewing. This makes it possible to develop large and complex questionnaires. To protect respondents from having to answer all these questions, routing structures and filter questions see to it that only relevant questions are asked, and irrelevant questions are skipped. It is not always easy to test the routing structure of these large and complex questionnaires. Sometimes, it helps us to make a flowchart of the questionnaire. Figure 3.19 contains a small example. It is the same questionnaire as in Figure 3.18.

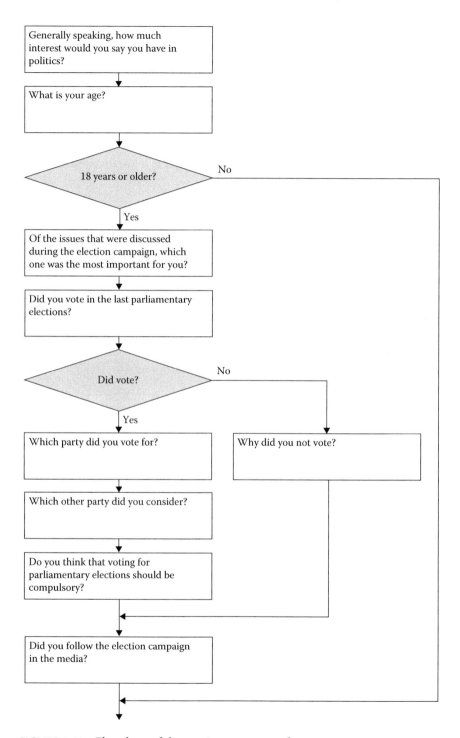

FIGURE 3.19  Flowchart of the routing structure of a questionnaire.

Testing a questionnaire may proceed in two phases. Converse and Presser (1986) suggest the first phase to consist of 25 to 75 interviews. The focus is on testing closed questions. The answer options must be clear and meaningful. All respondents must be able to find the proper answer. If the answer options do not cover all possibilities, there must be a way out by offering the special option *Other, please specify* ....

To collect the experiences of the interviewers in the first phase, Converse and Presser (1986) suggest letting them complete a small questionnaire with the following questions:

- Did any of the questions seem to make the respondents uncomfortable?

- Did you have to repeat any questions?

- Did the respondents misinterpret any questions?

- Which questions were the most difficult or awkward for you to ask? Have you come to dislike any questions? Why?

- Did any of the sections in the questionnaire seem to drag?

- Were there any sections in the questionnaire in which you felt that the respondents would have liked the opportunity to say more?

The first phase of questionnaire testing is a thorough search for essential errors. The second phase should be seen as a final rehearsal. The focus is no longer on repairing substantial errors, or on trying out a completely different approach. This is just the finishing touch. The questionnaire is tested in a real interview situation with real respondents. The respondents do not know they participate in a test. The number of respondents in the second phase is also 25 to 75. This is also the phase in which external experts could be consulted about the questionnaire.

## 3.7 SUMMARY

The questionnaire is a vital ingredient of a poll. To be able to assess the quality of a questionnaire, it should always be included in the poll report. If one cannot take a look at the questionnaire, it is not possible to determine the reliability of the answers.

The questionnaire is an instrument that measures facts, behavior, opinions, or attitudes. Questions must have been designed carefully, because even the smallest of changes in a question may cause substantial differences in the answers.

With respect to the text of a question, there are many caveats that must be avoided, such as unfamiliar wording, ambiguous question texts, long question texts, recall questions, leading questions, asking people things they do not know, sensitive questions, double questions, negative questions, and hypothetical questions.

There are several different questions types, such as open questions, closed questions, and numerical questions. Open questions should be avoided as much as possible. Closed questions should be preferred, either as closed question with one possible answer or as a check-all-that-apply question. The list of answer options should cover all possibilities. It should also not be too long.

Questions may suffer from satisficing. This is the phenomenon that respondents do not do all they can to provide the correct answer. Instead they give a more or less acceptable answer with minimal effort. Satisficing comes in many forms, like a primacy effect (preference for answer early in the list), a recency effect (preference for answer late in the list), preference for don't know or a neutral option, acquiescence (agreeing with the interviewer), or straight-lining (the same answer for all questions in a matrix of questions).

The order of the questions and the context of the questions are also important. Uninteresting and boring questions should be left to ask at the end of the questionnaire. Moreover, one should be aware that respondents may be framed by earlier questions in the questionnaire.

Every questionnaire should be tested before it is used in a poll, and poll reports should mention these tests. Large research organizations often have testing facilities and procedures. But smaller and less-experienced organizations may tend to skip testing because they do not have time and money to do this. It should be stressed that even the simplest test, like trying out the questionnaire on colleagues, can be very useful to detect major flaws.

# CHAPTER 4

# Data Collection

## 4.1 MODES OF DATA COLLECTION

The questionnaire of a poll must be filled in by all people selected in the sample. This can be done in various ways. These ways are called *modes of data collection*. Every mode has its advantages and disadvantages with respects to quality of the collected data and costs of the poll. Usually, the choice of the mode of data collection is a compromise between costs and quality.

To be able to assess the quality of the outcomes of a poll, it is important to know the mode of data collection used. As different modes of data collection may have a different impact on the collected data, this gives more insight in what will go right, and what may go wrong.

This chapter describes the main modes of data collection. The four main modes are as follows:

- *Face-to-face*: Interviewers visit the sample persons at home. The interviewers ask the questions, and the sample persons give the answers.

- *By telephone*: Interviewers call the sample persons by telephone. The interviewers ask the questions (by telephone), and the sample persons give the answers.

- *By mail*: Questionnaires are sent to the sample persons by mail. They are asked to complete the questionnaire and return it by mail to the researcher.

- *Online*: A link to the website with the (digital) questionnaire is sent to the sample persons. They are asked to go to poll website and fill in the questionnaire online.

The researcher has to take a decision in the setup phase of a poll about which mode of data collection to use. Two important aspects will probably determine this choice: quality and costs. Face-to-face interviewing and telephone interviewing make use of interviewers. Therefore, these modes of data collection are called *interviewer-assisted*. Interviewers can persuade people to participate in the poll, and they can also assist them in formulating the correct answers to the questions. Therefore, the collected data in interviewer-assisted polls tend to be of good quality. However, these modes of data collection are expensive because interviewers are expensive.

Mail and online polls do not employ interviewers. Therefore, these modes of data collection are called *self-administered*. The respondents fill in the questionnaire on their own. This may lead to *satisficing*. This is the phenomenon that respondents do not do all they can to provide the correct answer. Instead, they give a more or less acceptable answer with minimal effort. Consequently, the quality of the answers in a self-administered poll may not be very good. Of course, the costs of a mail poll or an online poll are much lower than that of a face-to-face or a telephone poll.

All data collection activities together are often called the *fieldwork*. In principle, this term refers to interviewers going into the field for a face-to-face poll. The term will, however, also be used for other modes of data collection.

The four main modes of data collection are mail polls, face-to-face polls, telephone polls, and online polls. This chapter compares these modes. Section 4.6 summarizes the most important differences between the modes. A more detailed treatment of various data collection modes can be found in, for example, De Leeuw (2008).

## 4.2 MAIL POLLS

*Mail interviewing* is the least expensive of the four data collection modes. The researcher sends a copy of the paper questionnaire to the people selected in the sample. They are invited to answer the questions and return the completed questionnaire by mail to the researcher. There are no interviewers. Therefore, it is a cheap mode of data collection. Data collection costs only involve mailing costs (letters, postage, and envelopes). A possible advantage of a mail poll can be that the absence of interviewers is experienced as less threatening for potential respondents. As a consequence, respondents are more inclined to answer sensitive questions correctly.

The absence of interviewers also has a number of disadvantages. They cannot provide additional explanation or assist the respondents in answering the questions. This may cause respondents to misinterpret questions or to give no answers at all. This has a negative impact on the quality of the collected data. Mail polls put high demands on the design of the paper questionnaire. It should always be clear to all respondents how to navigate through the questionnaire from one question to the next, and how to answer the questions.

As the persuasive power of the interviewers is absent, response rates of mail polls tend to be low. Of course, reminder letters can be sent, but this makes the poll more expensive, and it is often only partially successful. More often poll documents (questionnaires, letters, and reminders) end up in the pile of old newspapers.

In summary, the costs of a mail poll are relatively low, but a price must be paid in terms of quality: response rates are low, and also the quality of the collected data is not very good. However, Dillman et al. (2014) believe that good results can be obtained by applying his *Tailored Design Method*. This is a set of guidelines for designing and formatting questionnaires for mail polls. This method pays attention to all aspects of the data collection process that may affect response rates or data quality.

## 4.3 FACE-TO-FACE POLLS

A *face-to-face* poll is the most expensive of the three traditional data collection modes (face-to-face, telephone, and mail). The basic idea is that interviewers visit the sample persons at home. Well-trained interviewers will be successful in persuading reluctant persons to participate in the poll. Therefore, response rates of face-to-face polls are usually higher than those of mail polls. The interviewers can also assist respondents in giving the right answers the questions. This often results in data of better quality. Be aware that too much involvement should be avoided, as this may violate the principle that interviews must be as standardized as possible.

Presence of interviewers can also be a drawback when answering sensitive questions. Research suggests that respondents are more inclined to answer sensitive questions correctly if there are no interviewers in the room.

The interviewers can use show cards. *Show cards* are typically used for answering closed questions. Such a card contains the list of all possible answers to a question. It allows respondents to read through the list at their own pace and select the answer that reflects their situation or opinion.

The researcher may consider sending a letter announcing the visit of the interviewer. Such an *announcement letter* (also called an *advance letter* or a *prenotification letter*) can give additional information about the poll, explain why it is important to participate, and assure that the collected information is treated confidentially. As a result, the respondents are not taken by surprise by the interviewers. An announcement letter may also contain the telephone number of the interviewer. This makes it possible for the respondent to make an appointment for a more appropriate day and/or time. Of course, the respondent can also use this telephone number to cancel the interview.

Response rates of a face-to-face polls are a higher than those of mail polls, and the quality of the collected data is better. But a price has to be paid literally: face-to-face interviewing is much more expensive. A team of interviewers has to be trained and paid. Moreover, they have to travel a lot to visit the selected persons, and this costs time and money.

Face-to-face polls can be carried out in the traditional way (with a paper questionnaire), or in a computer-assisted way (with an electronic questionnaire). *Computer-assisted personal interviewing* (CAPI) became possible after the emergence of small portable computers in the 1980s. Interviewers take their laptop computer (or tablet) home of the respondents. There they start the interview program and record the answers to the questions.

Figure 4.1 shows an example of a screen of a CAPI poll. It was generated by the Blaise System. This system was developed by Statistics Netherlands. See, for example, Bethlehem (2009) for more information about Blaise.

The upper half of the screen contains the text of the current question to be answered (*Which party was your second choice?*). It is a closed question, which means that the respondent has to select and answer by clicking on one of the radio buttons.

The bottom half of the screen gives a condense view on the questionnaire. It shows which questions have already been answered (the first seven), and which questions still have to be answered. So the interviewers have an overview of where they are in the questionnaire. They can also see which answers have already been given.

The interviewer can only proceed to the next question after the current question has been answered correctly. If an error is made, the answer is not accepted. It must first be corrected. If necessary, the interviewer can go back to an earlier question and repair an earlier problem. The route through the questionnaire is determined by the system. So, only the correct route can be followed.

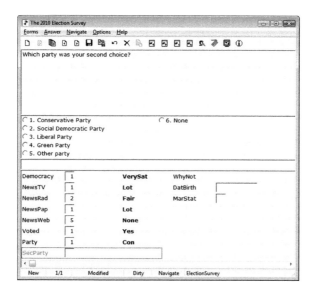

FIGURE 4.1    A question screen of the Blaise CAI system.

Application of computer-assisted interviewing for data collection has three major advantages:

- It simplifies the work of interviewers. They do not have to pay attention to choosing the correct route through the questionnaire. This is all taken care of by the interviewing software. Therefore, interviewers can concentrate more on asking questions and assisting respondents in getting the right answers.

- It improves the quality of the collected data, because answers can be checked and corrected by the software during the interview. If an error is detected, it is immediately reported on the screen, and interviewer and respondent together can correct the error. This is more effective than error treatment in traditional polls, where errors in paper forms can only be detected afterward, in the office of the researcher. The respondent is no longer available to correct the problem.

- Data are entered in the computer already during the interview, resulting in a clean record. So, no more subsequent data entry and data editing are necessary. This considerably reduces time needed to process the collected data and thus improves the timeliness of the poll results.

## 4.4 TELEPHONE POLLS

A *telephone poll* is somewhat less expensive than a face-to-face poll. Interviewers are also needed for this mode of data collection, but not as many as for face-to-face interviewing. They do not lose time traveling from one respondent to the next. They can remain in the call center of the research organization and conduct more interviews in the same amount of time. Moreover, there are no travel costs. Therefore, the costs of interviewers are less high. An advantage of telephone interviewing over face-to-face interviewing is that often respondents are more inclined to answer sensitive questions because the interviewer is not present in the room.

Telephone polls also have some drawbacks. Interviews cannot last too long, and questions may not be too complicated. Another drawback is often the lack of a proper sampling frame. Telephone directories may suffer from severe undercoverage because more and more people do not want their telephone number to be listed in the directory. Another development is that increasingly people replace their landline phone by a mobile phone. Mobile phone numbers are not listed in directories in many countries. As a result, there are serious coverage problems. For example, according to Cobben and Bethlehem (2005), only between 60% and 70% of the Dutch population can be reached through a telephone directory.

Telephone directories may also suffer from overcoverage. For example, if the target population of the poll consists of households, only telephone numbers of private addresses are required. Telephone numbers of companies must be ignored. It is not always clear whether a selected number refers to a private address or a company address (or both).

A way to avoid undercoverage problems of telephone directories is applying *random digit dialing* (RDD) to generate random telephone numbers. This comes down to using a computer algorithm for computing valid telephone numbers in a random fashion. One way to do this is by replacing the final digit of an existing number by a random different digit. Such an algorithm is able to generate both listed and unlisted numbers. It can also be used to generate mobile phone numbers. So, there is complete coverage of all people with a telephone. RDD also has drawbacks. In some countries, it is not clear what an unanswered number means. It can mean that the number is not in use. This is a case of overcoverage, for which no follow-up is needed. It can also mean that someone simply does not answer the phone, which is a case of nonresponse, which has to be followed up. Another drawback of RDD is that there is no information at all about nonrespondents. This makes correction for nonresponse very difficult. See also Chapter 9 (nonresponse).

The fast rise of the use of mobile phones has not made telephone interviewing easier. More and more landline phones are replaced by mobile phones. A landline phone is a means of contacting a household, whereas a mobile phone makes contact with an individual person. Therefore, the chances of contacting a member of the household are higher for landline phones. If persons can only be contacted through their mobile phones, it is often in a situation not fit for conducting an interview. Also, it was already mentioned that sampling frames in many countries do not contain mobile phone numbers. Note that people with both a landline phone and a mobile phone have a higher probability of being selected in the sample. So these people are over-represented in the sample. For more information about the use of mobile phones for interviewing, see, for example, Kuusela et al. (2006).

The choice of the mode of data collection is not any easy one. It is usually a compromise between quality and costs. In a large country like the United States, it is almost impossible to collect data by means of face-to-face interviewing. It requires so many interviewers that have to do so much traveling that the costs would be very high. Therefore, it is not surprising that telephone interviewing emerged in the United States as a major data collection mode. In a very small and densely populated country like The Netherlands, face-to-face interviewing is much more attractive. Coverage problems of telephone directories and low response rates also play a role in the choice for face-to-face interviewing.

Telephone polls can be carried out in the traditional way (with a paper questionnaire) or in a computer-assisted way (with an electronic questionnaire). The latter is called *computer-assisted telephone interviewing* (CATI). Interviewers in a CATI poll operate a computer running interview software. When instructed so by the software, they attempt to contact a selected person by telephone. If this is successful and the person is willing to participate in the survey, the interviewer starts the interviewing program. The first question appears on the screen. If this is answered correctly, the software proceeds to the next question on the route through the questionnaire, and so on.

Many CATI systems have a tool for *call management*. Its main function is to offer the right phone number at the right moment to the right interviewer. This is particularly important for cases in which the interviewer has made an appointment with a respondent for a specific time and date. Such a call management system also has facilities to deal with special situations like a busy number (try again after a short while) or no answer (try again later). This all helps one to increase the response rate as much as possible.

## 4.5 ONLINE POLLS

The rapid development of the internet in the 1990s led to a new mode of data collection: online polls. Some call it *computer-assisted web interviewing* (CAWI). The questionnaire is offered to the respondents through the internet. Therefore, such a poll is sometimes also called a *web poll.*

At first sight, online polls have a number of attractive properties. As so many people are connected to the internet, it is a simple means of getting access to a large group of potential respondents. Furthermore, questionnaires can be distributed at very low costs. No interviewers are needed, and there are no mailing and printing costs. Finally, polls can be launched very quickly. Little time is lost between the moment the questionnaire is ready and the start of the fieldwork. Consequently, it is a cheap and fast means to get access to a large group of people.

However, online polls also have some serious drawbacks. One possible drawback is undercoverage. If not all people in the target population of the poll have an internet connection, it is not possible to cover the whole population. Specific groups may be underrepresented or missing. This may lead biased estimates of population characteristics, and thus to wrong outcomes.

Another issue is sample selection. Preferably, one would like to have a sampling frame with e-mail addresses of all people in the target population. Often, such a sampling frame is not available, particularly if the target population is the general population. One way to avoid this problem is to use a different sampling frame. For example, if a list of mail addresses is available, the researcher could draw a sample of addresses and send a letter to the selected addresses with an invitation to participate in the poll. This letter also contains the link to the website with the online questionnaire. Note that this approach increases the costs of the poll (mailing costs). Another solution of the sample selection issue is to apply self-selection for recruiting respondents.

Because of the increasing popularity of online polls and the associated methodological problems, a special chapter is devoted to this mode of data collection, that is Chapter 10 (online polls).

## 4.6 THE CHOICE OF THE MODE OF DATA COLLECTION

Table 4.1 summarizes a number of advantages and disadvantages of the four main modes of data collection. This information can help us to assess the quality of a poll.

With respect to costs, the conclusion is clear: both face-to-face polls and telephone polls are expensive. The reason is that these polls use interviewers for data collection, and interviewers are expensive. For a mail poll, there are only printing and mailing costs. The costs of an online poll are similar if people are invited by mail to participate in the poll. The costs can even be less if respondents are recruited by means of self-selection. However, self-selection can be a source of systematic errors in the outcomes of the poll.

Coverage problems may arise if the sampling frame does not cover the target population. For a face-to-face poll, or a mail poll, an address list can be used. Usually, such a list provides good coverage, provided it is up-to-date. A very good sampling frame is a population register, but it is often not unavailable, or it is not accessible. The situation is problematic for a telephone poll. Telephone directories frequently suffer from substantial undercoverage. There are many unlisted landline phone numbers, and almost all mobile phone numbers will be missing. RDD may be used as an alternative, but this also has its problems. An online poll only has sufficient coverage if all people in the target population have internet access and an e-mail address. A random sample for such a poll can only be selected if there is a sampling frame containing these e-mail addresses. This is often not the case. Then, the researcher has to fall back on recruiting respondents by ordinary mail.

Nonresponse affects the outcomes of a poll. The more nonresponse there is, the larger the bias in the outcomes will be. There are many factors that determine the response rate of a poll. One important factor is whether the

TABLE 4.1    Advantages (+) and Disadvantages (−) of Various Modes of Data Collection

|  | Face-to-Face | Telephone | Online | Mail |
|---|:---:|:---:|:---:|:---:|
| Costs | − | − | + | + |
| Coverage | + | − | − | + |
| Nonresponse | + | + | − | − |
| Satisficing | − | − | − | − |
| Socially desirable answer | − | − | + | + |
| Checking | + | + | − |  |
| Routing | + | + | − |  |
| Timeliness | + | + | + | − |

poll is interviewer-assisted or not. Interviewer-assisted polls have higher response rates than self-administered polls. From the point of view of nonresponse, face-to-face polls have the best quality, followed by telephone polls, mail polls, and online polls.

The response rates in telephone polls are deteriorating rapidly. Pew Research Center (2012) reports that the responses rate of typical RDD telephone polls in the United States dropped from 36% in 1997 to 9% in 2012. According to Vavrek and Rivers (2008), most media polls (telephone polls using RDD) have response rates of around 20%. Response rates of RDD telephone polls also dropped in the United Kingdom to around 20%, and in inner cities, they are even down to 10%.

Satisficing is the phenomenon that respondents do not do all they can to provide the correct answers to the questions. Instead they give a more or less acceptable answer with minimal effort. Satisficing comes in many forms. It occurs in all types of polls. For example, self-administered polls may suffer from a primacy effect (preference for answer early in the list), and interviewer-assisted polls may be affected by a recency effect (preference for answer late in the list). Other examples are a preference for "don't know" or a neutral option, acquiescence (agreeing with the interviewer) in interviewer-assisted polls, and straight-lining (the same answer for all questions in a matrix of questions) in an online poll.

Sensitive questions are a problem. Such questions address topics which respondents may see as embarrassing. This may result in inaccurate answers. Respondents will avoid socially undesirable answers and give socially more acceptable answers. They may even give no answer at all. Examples of sensitive questions are questions about topics such as income, health, criminal behavior, or sexual behavior. As a rule, respondents answer sensitive more honestly in a mail poll or an online poll. This is because there are no interviewers present. Respondents, who are on their own, feel more free in giving the true answer.

Completed questionnaires may contain errors. Wrong answers may be caused by respondents not knowing the right answers or not understanding the questions. They may also be caused by interviewers recording wrong answers. Therefore, it is important to check completed questionnaires for errors. This is possible if data are collected by means of computer-assisted interviewing. The interviewing software can have built-in checks. Detected errors are reported immediately, after which the problem can be solved, because the respondents are still there. Checking paper questionnaires is less effective, because it has to be done afterward, so that respondents are not available anymore. The conclusion is that the computer-assisted versions of face-to-face polls and telephone polls produce the best data. Checks can be included in online polls, but often they are omitted. This because some researchers think they should not bother respondents with error messages. They fear this may lead to nonresponse. The price they pay is data of less quality.

Many questionnaires contain routing instructions. These instructions guide the respondents through the questionnaire. They force respondents to answer relevant questions and skip irrelevant questions. Routing instructions also reduce the number of questions that have to be answered. This may have a positive effect on the response rate of the poll. So routing instructions lead to better data. They can be implemented in CAPI, CATI, and online polls. Not every online poll has routing instructions, however. Sometimes, they are omitted. The idea is to let respondents completely free in choosing the questions to answer. This could increase the response rate, but it also leads to more incomplete questionnaires. Routing instructions can also be included in paper questionnaire, but respondents cannot be forced to follow these instructions. Nothing prevents them from ignoring some relevant questions, and to jump to questions that are not on the route. So the quality of the collected data is less good.

Timeliness of a poll is important. Researchers want the results of their polls as soon as possible, without any delay. From this point of view, quick polls are the best polls. An online poll with recruitment based on self-selection is the quickest poll. There are examples of such polls that were carried out in one day. The slowest poll probably is a mail poll, or an online poll with recruitment by mail. Invitation letters are sent by ordinary mail, after which the researcher has to wait for a response. This can take a while. An attempt can be made it to speed up the data collection process by sending reminders. Unfortunately, the response will often remain low. In the case of a telephone poll, the researcher has more control over the fieldwork. If sufficient interviewers are available, the fieldwork can be completed in a few days. Note that recontact attempts in the case of no-answer can prolong the fieldwork. The fieldwork of a face-to-face poll takes longer than that of a telephone poll, particularly recontact attempts in case people were not at home. This is because interviewers have to travel a lot. Nevertheless, the fieldwork can be completed in a limited number of days if a sufficient number of interviewers are deployed, and they are successful in getting cooperation.

## 4.7 SUMMARY

The four most used modes of data collection are face-to-face polls, telephone polls, mail polls, and online polls. These modes of data collection differ with respect to the quality of the data obtained and the costs of the poll. The choice of the mode is usually a compromise between quality and costs.

From the point of view of quality, a CAPI poll is the best choice. Unfortunately, CAPI polls are expensive. This is probably the reason that many poll organizations gradually replace CAPI polls by online polls. If it is important to keep costs within limits, and there are no high demands for data quality, it could be better to carry out an online poll.

Telephone polls have always been fairly popular. A drawback was the lack of availability of sampling frames with good coverage of the population. The fast rise of the use of mobile phones has complicated the situation even further. RDD could be an alternative way of selecting a sample. A further complication is that the response rates of these types of polls are lower and lower. They tend to be not higher than 20%. Therefore, it is not unlikely that researchers may cease to conduct telephone polls.

Mail polls are a cheap mode of data collection. Unfortunately, response rates are often low, and data quality is low. It is also a slow mode of data collection. If costs are an important issue, it may be better to do an online poll instead of a mail poll.

# CHAPTER 5

# Sampling

## 5.1 BY A SAMPLE WE MAY JUDGE THE WHOLE PIECE

*By a sample, we may judge the whole piece.* This is a well-known quote from the English translation of the famous book *Don Quixote* by the Spanish writer Miguel de Cervantes (1547–1616). It is just one example of a method that is probably as old as mankind. Other examples are a cook in the kitchen taking a spoonful of soup to determine its taste, and a brewer only needing one sip of beer to test its quality.

Similar examples are encountered when tracing the roots of the Dutch word for sample, which is *steekproef*. The origin of this word is not completely clear. Some think it is a translation of the German word *Stichprobe*, which is a combination of *Stich* (dig, stab, and cut) and *Probe* (test and try). Already in 1583, the word was used to describe a technique in mining. With a kind of spoon a small amount was taken from a melted substance to determine the quantity of metal in it. Others believe that *steekproef* goes back to cheese making. Cheese making in The Netherlands started already in prehistoric times. There were already cheese markets in the Middles Ages. Cheese masters cut (*steken* = to cut) a sample from the cheese and tasted (*proeven* = to taste) its quality.

The English term *sample* comes from the old French *essample*. This has its roots in the Latin *exemplum*, which literary means *what is taken out*.

Apparently, the idea of sampling is already very old. So why not use it for investigating a population of people? Why not apply it in polls? Figure 5.1 shows how sampling should be used in polls. The first step is to select a sample from the population. The people in the sample complete a

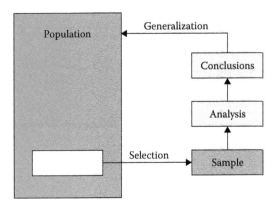

FIGURE 5.1   The concept of sampling.

questionnaire, and so the sample data are collected. The data are analyzed, and conclusions are drawn. Finally, the conclusions are generalized from the sample to the population.

The generalization from the sample to the population is a critical step in this process. The question is whether poll results can always be translated from the sample to the population. Yes, it is possible, but not always. It all depends on the way the sample is selected. If this is properly done, valid conclusions can be drawn about the state of the population. And if the sample is not selected properly, there is a serious risk of drawing wrong conclusions. So be careful with the results of a poll. First, check whether the sample was selected properly.

What is *properly*? The message in this chapter is that a proper sample is a random sample. The sample must have been drawn by means of probability sampling. It will be explained in detail what a random sample is, so that it can be recognized when considering poll results. It will also be shown what the problems are if no probability sampling is used.

Random samples come in many forms. Three forms are described in this chapter. The most frequency used one is the *simple random sample*. This is a form of probability sampling in which every person in the target population has the same probability of selection. *Systematic sampling* is an easier to implement approximation of simple random sampling. Furthermore, *two-stage sampling* is applied if a list of addresses is available, but individuals must be selected. This results in random sampling with unequal selection probabilities. There are more forms of random sampling, like stratified sampling and cluster sampling. They are beyond the scope of this book. For more information, see Bethlehem (2009).

Sometimes publications about polls only mention that the sample was representative. What does this mean? Is the sample OK, or is it a cover-up for a bad sample? This chapter explains what the relationship is between a random sample and a representative sample.

## 5.2  A REPRESENTATIVE SAMPLE?

For a period of almost 40 years, from 1895 to 1934, there was intensive discussion about how to select a sample from a population. In the end, experts like Bowley (1906) and Neyman (1934) made clear that by far the best way of doing this is to apply probability sampling. If a random sample is drawn, it is always possible to compute valid (unbiased) estimates of population characteristics. Moreover, by computing their margin of error it can also be made clear how good the estimates are.

The paradigm of probability sampling has been in use in social research, official statistics, and market research for many years now. And it works well, as it allows for drawing well-founded, reliable, and valid conclusions about the state of the population that is being investigated.

It is often said that a sample is a good sample if it is a representative sample. Unfortunately, it is usually not clear what *representative* means. Kruskal and Mosteller (1979a–c) explored the literature and found nine different meanings of *representative*:

1. *General acclaim for data.* This means not much more than a general assurance, without evidence, that the data are OK. This use of *representative* can typically (without further explanation) be found in the media.

2. *Absence of selective forces.* The selection process did not favor some people, either consciously or unconsciously.

3. *Miniature of the population.* The sample is a scale model of the population. The sample has the same characteristics as the population. The sample proportions are in all respects similar to population proportions.

4. *Typical or ideal cases.* The sample consists of people that are *typical* of the population. These people are *representative elements.* This meaning probably goes back to the idea of *l'homme moyenne* (*average man*) that was introduced by the Dutch/Belgian statistician Quetelet (1835, 1846).

5. *Coverage of the population's heterogeneity.* The variation that exists in the population must also be found back in the sample. This means, for example, that the sample should also contain atypical elements.

6. *A vague term, to be made precise.* Initially the term is simply used without describing what it is. Later on, it is explained what is meant by it.

7. *A specific sampling method has been applied.* A form of probability sampling has been applied giving equal selection probabilities to each person in the population.

8. *Permitting good estimation.* All characteristics of the population and the variability must be found back in the sample, so that it is possible to compute reliable estimates of population parameters.

9. *Good enough for a particular purpose.* Any sample that shows that a phenomenon thought to be very rare or absent occurs with some frequency will do.

It is clear that use of the term *representative* can be confusing. Therefore, it is better not to use this term unless it is explained what it means. In this book, a sample is called *representative* if it was selected by means of probability sampling, and each person in the population had the same probability of selection. Such a sample is called a *simple random sample*. It corresponds to meaning 7 in the list of Kruskal and Mosteller.

This book sometimes also uses the notion representative with respect to a variable. A sample is *representative with respect to a variable* if the distribution of the variable in the sample is equal to the distribution in the population. For example, according to the U.S. Census Bureau, the U.S. population consists of 49% males and 51% females. So the sample is representative with respect to gender if the percentages of males and females in the sample are 49% and 51%. Representativity with respect to a set of variables corresponds to meaning 3 of Kruskal and Mosteller.

Note that a representative sample does not imply representativity with respect to all variables measured in the poll. One can say, however, that a representative sample is approximately representative with respect to all variables. So, again as an example, the percentages of males and females in a representative sample will be more or less equal to the percentages of males and females in the population.

So a poll should be based on a representative sample, in which representative means that people are selected at random and with equal probabilities. But how are random samples drawn in practice? This is the topic of this chapter.

## 5.3 THE SAMPLING FRAME

How is a sample drawn from a target population? How are a number of people selected that can be considered representative? A researcher needs a sampling frame to do this. A *sampling frame* is a list of all people in the target population. And it must be clear for all persons in this list how to contact them.

The choice of the sampling frame depends on the mode of data collection. For a face-to-face poll or a mail poll a list of addresses is required. For a telephone poll, the researcher must have a list of telephone numbers. And for an online poll, he must have a list of e-mail addresses.

A sampling frame can exist on paper (e.g., a card-index box for the members of a club or a telephone directory) or in a computer (e.g., a database containing names and addresses or a population register). If such lists are not available, detailed geographical maps are sometimes used to locate people.

The sampling frame should be an accurate representation of the population. There is a risk of drawing wrong conclusions from a poll if the sample is selected from a sampling frame that differs from the population. Figure 5.2 shows what can go wrong.

The first problem is *undercoverage*. This occurs if the target population contains people that have no counterpart in the sampling frame. Such

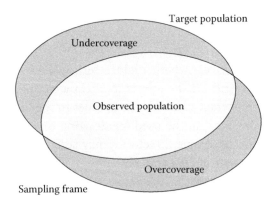

FIGURE 5.2   Coverage problems in a sampling frame.

persons can never be selected in the sample. An example of undercoverage is a poll in which the sample is selected from a population register. Illegal immigrants are part of the population but will never be encountered in the sampling frame. Another example is an online poll, in which respondents are selected via the internet. There will be undercoverage due to people without internet access. Undercoverage can have serious consequences. If people outside the sampling frame systematically differ from those in the sampling frame, estimates of population characteristics may be seriously biased. A complicating factor is that it is often not very easy to detect undercoverage.

Another problem that can occur in a sampling frame problem is overcoverage. This refers to the situation in which the sampling frame contains people that do not belong to the target population. If such people end up in the sample and their data are used in the analysis, estimates of population characteristics may be affected. It should be rather simple to detect overcoverage as these people should experience difficulties to answer the questions in the questionnaire.

Some countries, such as The Netherlands and the Scandinavian countries, have a *population register*. Such a register contains all permanent residents in the country. It contains for each person name and address and also the values of other variables, such as date of birth, gender, marital status, and country of birth. Such population registers are an ideal sampling frame, because they usually have little coverage problems.

Another frequently used sampling frame is a *postal address file* (PAF). Postal service agencies in several countries maintain databases of all delivery points in the country. Examples are The Netherlands, the United Kingdom, Australia, and New Zealand. Such databases contain postal addresses of both private houses and companies. Typically, a PAF can be used to draw a sample of addresses, and therefore also of households.

It is sometimes not clear whether addresses in a PAF belong to private houses or to companies. If the aim is to select a sample of households, there may be overcoverage caused by companies in the file.

A *telephone directory* can be used for drawing a sample of telephone numbers. However, telephone directories may suffer from serious coverage problems. Undercoverage occurs because many people have unlisted numbers, and some will have no telephone at all. Moreover, there is a rapid increase of the use of mobile phones. In many countries, mobile phone numbers are not listed in directories. This means that young people with only a mobile phone are missing in the sampling frame and therefore may be seriously underrepresented in a poll. A telephone directory also suffers from overcoverage, because it contains telephone numbers of shops, companies, and so on. Hence, it may happen that people are contacted that do not belong to the target population. Moreover, some people may have a higher than assumed contact probability, because they can be contacted both at home and in the office.

Undercoverage problems of telephone directories can be avoided by applying *random digit dialing* (RDD). It means that a computer algorithm is used to generate valid random telephone numbers. One way to do this is taking an existing number from the telephone directory and to replace its final digit by a random other digit. An RDD algorithm can produce both listed and unlisted numbers, including numbers of mobile phones. So there is complete coverage.

RDD also has its drawbacks. In some countries, it is not clear what an unanswered number means. Maybe the number is not in use. Then it is a case of overcoverage. No follow-up is needed. It can also mean that someone simply did not answer the phone. This is a case of nonresponse, which has to be followed up. Another drawback of RDD is that there is no information at all about nonrespondents. This makes correction for nonresponse very difficult. See Chapter 6 for a detailed treatment of the topic of nonresponse.

For conducting an online poll, sampling frame with e-mail addresses is required. Unfortunately, for many populations, such sampling frames are not available. Exceptions are a poll among students of a university (where each student has a university supplied e-mail address) or a poll among employees of a large company (where every employee has a company e-mail address).

If you intend to use the results of a poll, it is important to check which sampling frame was used to select the sample. You have to assure yourself that the sampling frame exactly covered the population. If this not the case, the poll may miss a significant part of the population. In fact, a different population is investigated, and this is the population as far as it is covered by the sampling frame.

## 5.4 HOW NOT TO SELECT A SAMPLE

This chapter stresses the importance of selecting a random sample. Unfortunately, there are also other ways of selecting a sample that do not result in a representative sample. Problems are often caused by the lack of proper sampling frames. Some examples of bad polls are given in this section. If you see a poll that resembles one of these examples, be careful. It might be an unreliable one.

## 5.4.1 A Poll in a Shopping Mall

A local radio station conducted a radio-listening poll. To quickly collect a lot of data, it was decided to send interviewers to the local shopping mall on Saturday afternoon. So this is a face-to-face poll. There were a lot of people there. So it was not too difficult to get a large number of completed questionnaires. Analysis of the collected data led to a surprising conclusion: almost no one listened to the sports program that was broadcasted every Saturday afternoon.

Of course, this conclusion was not so surprising. The only people interviewed were those in the shopping center at Saturday afternoon. Moreover, they were not listening to the sports program on the radio. In fact, this approach reduced the target population of the poll from all inhabitants to only those shopping on Saturday. So, the sample was not representative.

## 5.4.2 A Poll in a Magazine

A publisher distributes a free door-to-door magazine in a town each week. The editors of the magazine wanted to know how many inhabitants really read the magazine. So, they decided the carry out a poll. In a specific week, they included a questionnaire form in the magazine. People were asked to complete the questionnaire and to return it to the publisher. Of course, one of the most important questions was whether the magazine was read. The returned forms showed that everyone read the magazine. The publisher was happy.

However, the publisher did not realize that there was something wrong. There were a lot of people who did not read the magazine and threw it away immediately. They did not encounter the questionnaire form, and therefore, they could not answer the questions. In fact, this sampling approach restricted the target population to only those reading the magazine.

## 5.4.3 A Poll about Singles

Problems can occur in a poll if the units in the sampling frame are different from those in the target population. Suppose a researcher wants to estimate the percentage of single people in a population. His idea is to draw a simple random sample of people from the population. He only has a PAF. So he can only select a sample of households. He draws a random person (only taking into account persons who belong to the target population) in each selected household.

If the researcher treats this sample as a simple random sample, he makes a mistake. The problem is that selection probabilities depend on the size of the household. Someone in a single household has a higher selection probability than someone in a large household. Consequently, single people will be overrepresented in the sample, and therefore, the estimate of the percentage of singles will be too large.

### 5.4.4 A Household Poll

The situation is the other way around if a researcher wants to select a simple random sample of households, and he only has available a sampling frame containing persons. A population register is an example. Now large households have a larger selection probability than single-person households, because larger households have more people in the sampling frame. In fact, the selection probability of a household is proportional to the size of the family. Therefore, the sample will not be representative.

## 5.5 RANDOM NUMBERS

To conduct a poll, a researcher needs to have a random sample. How to draw a random sample? To select a random sample from the sampling frame, a random number generator is required. Some tools for generating random numbers are described in this section. These random numbers form the basis for drawing a random sample.

Let us assume that there is a sampling frame for the population to be investigated. Also suppose the people in the sampling frame are numbered 1, 2, 3, …, $N$, where $N$ is the size of the population. To select a random sample from this population, random numbers have to be generated in the range from 1 up to and including $N$. How to generate these random numbers?

A first idea could be to let a person pick some arbitrary numbers. Unfortunately, this does not result in a random sample. People seem not able to pick numbers in such a way that all numbers have the same probability of selection. This is clear from an experiment in which a sample of 413 persons was asked to pick an arbitrary number in the range from 1 up to and including 9. The results are summarized in Figure 5.3.

If people would pick numbers completely at random, each of the nine numbers should have been mentioned with approximately the same

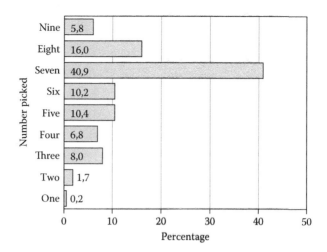

FIGURE 5.3   Picking a random number.

frequency of 100/9 = 11%. This is definitely not the case. Figure 5.3 shows that many people seem to have a high preference for the number *seven*. More than 40% mentioned *seven*. Apparently *seven* is more popular than any other number. The numbers *one* and *two* are almost never mentioned. The conclusion must be that people cannot select a random sample.

What is needed is an objective probability mechanism that guarantees that every person in the population has exactly the same probability of being selected. Such a mechanism is called a *random number generator*. It is a device (electronic or mechanical) with the following properties:

- It can be used repeatedly.
- It has $N$ possible outcomes that are numbered 1, 2, ..., $N$, where $N$ is known.
- Each time it is activated, it produces one of the $N$ possible outcomes.
- Each time it is activated, all possible outcomes are equally probable.

The main property of a random number generator is that its outcome is completely unpredictable. All prediction methods, with or without knowledge or use of past results, are equivalent.

The perfect random number generator does not exist in practice. There are, however, devices that come close to it. They serve their purpose as a random number generator. The proof of the pudding is in the eating:

FIGURE 5.4 Example of a random number generator: dice. (Printed with permission of Imre Kortbeek.)

the people living in the princedom of Monaco do not pay taxes as the random number generators in the casino of Monaco provide sufficient income for the princedom.

A coin is a simple example of a random number generator. The two outcomes *heads* and *tails* are equally probable. Another example of a random number generator is a dice, see Figure 5.4. Each of the numbers 1 to 6 has the same probability, provided the dice is *fair*.

A coin can only be used to draw a sample from a population of two elements. And dice can be used only for a population of six elements. This is not very realistic. Target populations are usually much larger than that. Suppose, a sample must be drawn from a target population of size 750. A coin or a dice cannot be used. What can be used is a 20-sided dice (see Figure 5.5).

FIGURE 5.5 A 20-sided dice. (Printed with permission of John Wiley & Sons from Bethlehem 2009.)

FIGURE 5.6    A calculator with a random value function.

Such a dice contains the numbers from 1 to 20. If 10 is subtracted from the outcomes 10 and higher, and 10 is interpreted as 0, the dice contains twice the numbers from 0 to 9. Three throws of this dice produce a three-digit number in the range from 0 to 999. If the outcome 0 and all outcomes over 750 are ignored, a random number is obtained in the range from 1 up to and including 750.

In practice, larger samples are drawn from much larger population, like a sample of 1000 people from the target population of all potential voters in a country. It is not realistic to use dice for this. A more practical approach is to use a calculator, a spreadsheet program, or some other computer software.

Some calculators, see, for example, Figure 5.6, have a function to generate *random values* from the interval [0, 1). Every value between 0 and 1 is possible. The value 0 can occur, but not the value 1. For example, the CASIO FX-82 calculator has a button RAN#. Each time this button is pressed, a new random value from the interval [0, 1) appears on the display.

For a random sample from a population, random numbers are needed in the range from 1 to $N$, where $N$ is the size of the population. At first sight, random values from the interval [0, 1) seem not useful for this. Fortunately, there is a simple procedure to transform random values from [0, 1) into random numbers from 1 to $N$. See Table 5.1. In the example, the population size is assumed to be 18,000 ($N = 18,000$).

The researcher could also have used a spreadsheet program to generate random numbers. Here it is shown how to do this with MS Excel. First, fill

TABLE 5.1  Generating a Random Number with a Calculator

| Step | Action | Example |
|------|--------|---------|
| 1 | Draw a random value RAN from the interval [0, 1). | 0.268 |
| 2 | Multiply this value by the size of the target population N. This produces a value in the interval [0, N). | 4824.000 |
| 3 | Remove the decimal part of the result. This produces an integer number in the range from 0 to N − 1. | 4824 |
| 4 | Add 1 to this integer number. This produces an integer number in the range from 1 to N. | 4825 |

column A with random values from [0, 1). This can be done with the function RAND(). Next, random numbers are created in column B by applying the procedure in Table 5.1.

Suppose, the population size is 18,000 and 10 random numbers are required. First, the cells A1, A2, ..., A10 are filled with random values using RAND(). Next, random numbers are computed in cells B1, B2, ..., B10 with the computations $= 1 + INT(A1 \times 18,000)$, $= 1 + INT(A2 \times 18,000)$, and so on. Figure 5.7 shows the result of this process.

There are many more software tools for generating random numbers; such tools can be found on the internet. One example is the website www. random-sample.nl. First, the lower bound (usually 1) and the upper bound (the population size N) must be entered. Then a random number will be produced every time the button Draw is clicked. See Figure 5.8 for an example.

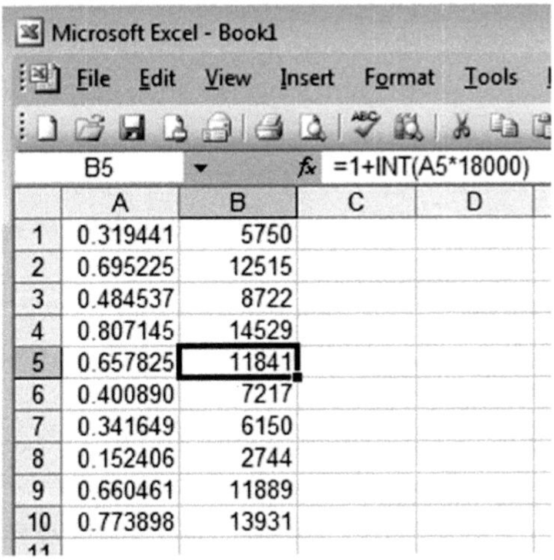

FIGURE 5.7  Generating random numbers in a spreadsheet.

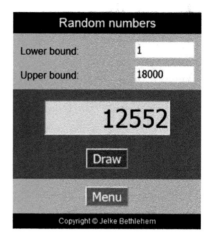

FIGURE 5.8 Generating random numbers with a web tool.

## 5.6 SIMPLE RANDOM SAMPLING

The obvious way to select a probability sample is to assign the same probability of selection to each person in the target population. This is called *simple random sampling*. This is the most frequently used sampling design. It is, however, possible to use other sampling designs, in which the selection probabilities are not the same. Some of them will be briefly discussed later in this chapter.

If a dice is thrown a number of times, it is possible that a certain number appears more than once. The same applies to other random number generators: if a sequence of random numbers is generated, some numbers may occur several times. The consequence would be that the corresponding person is selected more than once in the sample. Would this mean interviewing this person twice? Or should his questionnaire be copied? This is all not very meaningful. Therefore, *sampling without replacement* is preferred. This is a way of sampling in which each person can appear at most only once in a sample.

A lotto machine is a good example of sampling without replacement (see Figure 5.9). A selected ball is not replaced in the population. Therefore, it cannot be selected a second time.

The procedure for selecting a sample without replacement is straightforward: generate a sequence of random numbers in the range from 1 to $N$ using some random number generator. If a number is produced that was already generated previously, it is simply ignored. Continue this process until the sample size is reached. The process is summarized in Table 5.2.

FIGURE 5.9    A lotto machine: selecting a sample without replacement. (Printed with permission of Imre Kortbeek.)

TABLE 5.2    Generating a Sample without Replacement with a Random Number Generator

| Step | Action |
|---|---|
| 1 | Draw a random value $RAN$ from the interval $[0, 1)$. |
| 2 | Multiply this value by the size of the target population $N$. This produces a value in the interval $[0, N)$. |
| 3 | Remove the decimal part of the value. This produces an integer number in the range from 0 to $N - 1$. |
| 4 | Add 1 to this integer number. This produces an integer number in the range from 1 to $N$. |
| 5 | If this number is already in the sample, ignore it, go back step 1, and make a new attempt. If the number is new, add it to the sample. |
| 6 | If the sample size has not been reached, go back to step 1, and generate another number. |

For a bigger sample, the manual procedure in Table 5.2 may be too cumbersome. An alternative is to draw the sample with a spreadsheet program. Table 5.3 describes how to do this with MS Excel.

For much large samples of, say, of few thousand objects, from very large target populations, the spreadsheet approach is still cumbersome. It may be better to develop special software for this situation. There are also websites that can generate simple random samples. One such website is www.random-sample.nl. See Figure 5.10 for an example.

TABLE 5.3   Generating a Simple Random Sample without Replacement with a
Spreadsheet

| Step | Action |
|------|--------|
| 1 | Fill column *A* with the sequence numbers of the persons in the target population. These are the numbers from 1 to *N*, where *N* is the size of the population. The function *ROW()* can be used for this. |
| 2 | Fill column *B* with random values from the interval [0, 1). The function *RAND()* can be used for this. The spreadsheet fragment below on the left shows an example. |
| 3 | Select *Options* in the *Tools* menu and open the tab *Calculation*. Set *Calculation* to *Manual*. |
| 4 | Select columns *A* and *B*, order this block on column *B*. The result is something like in the spreadsheet fragment below on the right. |
| 5 | The sample consists of the set of numbers in the first part of column *A*. If, for example, a sample of 10 persons is required, take the first 10 numbers. |

*After step 2:*

| | A | B |
|----|----|--------|
| 1 | 1 | 0.677108 |
| 2 | 2 | 0.851208 |
| 3 | 3 | 0.902135 |
| 4 | 4 | 0.624257 |
| 5 | 5 | 0.914697 |
| 6 | 6 | 0.276631 |
| 7 | 7 | 0.880277 |
| 8 | 8 | 0.255902 |
| 9 | 9 | 0.543253 |
| 10 | 10 | 0.785882 |
| 11 | 11 | 0.862632 |
| 12 | 12 | 0.910895 |
| 13 | 13 | 0.490071 |

*After step 4:*

| | A | B |
|----|-----|----------|
| 1 | 903 | 0.000018 |
| 2 | 474 | 0.000964 |
| 3 | 286 | 0.001410 |
| 4 | 955 | 0.003571 |
| 5 | 982 | 0.003581 |
| 6 | 595 | 0.003699 |
| 7 | 399 | 0.004868 |
| 8 | 790 | 0.006207 |
| 9 | 689 | 0.006324 |
| 10 | 422 | 0.007093 |
| 11 | 292 | 0.007166 |
| 12 | 762 | 0.007327 |
| 13 | 519 | 0.007497 |

FIGURE 5.10   Selecting a sample without replacement with a web tool.

The example shows the result of drawing a simple random sample of size 50 from a target population of size 18,000. Note that the web tool also arranges the sample numbers in increasing order.

## 5.7  SYSTEMATIC SAMPLING

Nowadays, most sampling frames are digital lists that are generated from databases in computers. Sometimes, however, sampling frames still exist in physical form. One example is a rolodex, a small desktop file containing paper cards for names, addresses, and telephone numbers. Another example is a paper version of a telephone directory.

In some countries, it is possible to use population registers as sampling frames. In the early days, these registers consisted of a system of personal paper cards. See Figure 5.11 for an example. It shows the population register as it existed in The Netherlands from 1934 to 1994. Each municipality had its own population register. These registers were dynamic in the sense that day-to-day changes in the population were recorded in it. Therefore, they were up-to-date.

Manually selecting a simple random sample from a long physical list can be a lot of work. Particularly, if the people in the sampling frame are not

FIGURE 5.11  A population register in The Netherlands (1946). (Courtesy of Henk Blansjaar, November 8, 1946.)

numbered, it is not so easy to find, say, person 1341. For such situations, *systematic sampling* can be considered as an alternative for simple random sampling.

The first step in selecting a systematic sample is to determine the number of people in the target population (the *population size N*), and the number of people in the sample (the *sample size n*). It is assumed that dividing the population size by the sample size results in an integer number. If this is not the case, some modifications are required. See, for example, Bethlehem (2009) for more details about this.

The population size $N$ and the sample size $n$ are used to compute the step length $S$. The *step length* is the length of the jump used to jump through the sampling frame. The step length is obtained by dividing $N$ by $n$: $S = N/n$.

The next step in selecting a systematic sample is to compute the starting point $B$. The *starting point* is the sequence number of the first person in the sample. To obtain the starting point, a random number from the range of numbers from 1 up to and including the step length $S$ must be drawn.

Together, the starting point $B$ and the step length $S$ fix the sample. The first person in the sample is obtained by taking the starting point. The next person is obtained by adding the step length to the starting point. The process of adding the step length is continued until the end of the sampling frame is reached. The sample selection process is summarized in Table 5.4.

Here is an example of drawing systematic sample. Suppose, a researcher wants to conduct a radio-listening poll in the town of Harewood. There is a list with all 9500 addresses of households in the town. A systematic sample of 500 addresses has to be selected. Then, the *step length* is equal to $9500/500 = 19$.

TABLE 5.4    Generating a Sample without Replacement with a Random Number Generator

| Step | Action |
|---|---|
| 1 | Determine the population size $N$ and the sample size $n$. |
| 2 | Compute the step length $S$ by dividing the population size $N$ by the sample size $n$: $S = N/n$. |
| 3 | Compute the starting point $B$. Draw a random number from the range 1, 2, ..., $S$. This is the first person in the sample. |
| 4 | Add the step length $S$ to the starting point $B$. This is the next person in the sample. |
| 5 | If the end of the sampling frame has not yet been reached, go back to step 4. |
| 6 | The resulting sample exists of the persons with sequence numbers $B, B + S$. $B + 2 \times S, B + 3 \times S, ..., B + (n-1) \times S$. |

Next, the starting point must be drawn from the integer numbers in the range from 1 to 19. Suppose, the number 5 is drawn. Then this is the sequence number of the first household in the sample. All other sequence numbers in the sample are obtained by repeatedly adding the step length 19. The result is 5, 24, 43, ..., 9448, 9567, and 9486.

A warning is in place if a systematic sample is drawn from a sampling frame. Systematic sampling assumes that the order of the persons in the sampling frame is completely arbitrary. There should be no relation between the order of the persons and the variables that are measured in the poll.

A simple example illustrates the danger of selecting a systematic sample. Suppose, a researcher wants to conduct a poll about living conditions in a newly built neighborhood. All streets have 20 houses, numbered from 1 to 20. The sampling frame is ordered by street and house number. Suppose, he draws a systematic sample with a step length of 20. This means he selects one house in each street. Then there are two possibilities: (1) if the starting point is 1, he only has corner houses in his sample and (2) if the starting point is unequal to 1, he has no corner houses at all in his sample. So each possible sample is far from representative (with respect to housing conditions and related variables). Either there are too many or too few corner houses in the sample. This may affect the validity of the conclusions drawn from the poll.

If a systematic sample is selected from a list sorted by street name or postal code or from a telephone directory, it is often not unreasonable to assume there is no relationship between the order of the people in the list and the target variables of the survey. Then a systematic sample is more or less representative, and can it be considered similar to a simple random sample.

## 5.8 TWO-STAGE SAMPLING

For sampling households, there should be a sampling frame consisting of households. And for drawing a sample of persons, the sampling frame should be a list of persons. Unfortunately, the proper sampling frame is not always available in practice. If the target population of a poll is the general population, it would be nice to have a population register available. Often this does not exist, or it is not accessible. An alternative could be to use an address list. Selecting a sample of persons from a list of addresses is a two-step process: first addresses are drawn, and then one or more persons are drawn at each selected address. This called a *two-stage sample*.

A question that comes up is how many persons to draw at each selected address. One approach is to take all persons at the address who belong to the target population. Another approach is to take just one person at each selected address. In many situations, it is not very meaningful to interview several persons at one address. Often behavior, attitudes and opinions of persons at the same address are more or less similar. If this is the case, interviewing more persons at an address will not provide more information. It is more effective to spread the sample over more addresses and interview one person per address.

So, the advice will often be to draw one person at each selected address. How to do this? The first step would be to make a list of all persons at the selected address (as far as they belong to the target population). For an election poll, for example, this could be all persons living at the address and having an age of 18 years or older. One person must be selected randomly from this list. If the poll is conducted face-to-face or by telephone, interviewers have to do this selection. Moreover, if the poll is a mail poll or online poll, one of the members of the household has to do it. To keep things simple, the selection procedure must be easy. Often, the *first birthday procedure* is applied: select the person which the next birthday. This assumes that there is no relationship between the date of someone's birthday and the topic of the poll. Another way the select a person is to use a *Kish grid*. It uses a preassigned table of random numbers to find the person to be interviewed. For details, see Kish (1949).

One has to realize that by drawing one person at each selected address, the selection probabilities of persons are not equal anymore. Assuming addresses correspond to households (disregarding that it now and then can happen that there are more households living at the same address), each household has the same probability of being selected in the sample. But persons in a large household have a smaller selection probability than persons in small households. If estimates of population characteristics are computed, the researcher must take this into account by correcting for the unequal selection probabilities. If this is omitted, wrong conclusions may be drawn from the poll, because persons from large households are underrepresented. Therefore, the sample is not representative. Note that to be able to correct for the unequal selection probabilities, the number of household members (as far as they belong to the target population) of each selected household must be recorded.

## 5.9  QUOTA SAMPLING

The importance of probability sampling was made clear in the previous sections. If a random sample is selected, valid (unbiased) estimates of population characteristics can be computed. And also, the precision of the estimates can be determined.

There are, however, other ways of selecting a sample. Together, these techniques are called *nonprobability sampling*. The big disadvantage of these techniques is that there is no guarantee they lead to good estimates of population characteristics. So, one must be careful when poll results are based on nonprobability samples.

In this section and Section 5.10, two familiar types of nonprobability sampling are described. This section is about *quota sampling*, and Section 5.10 is about *self-selection sampling* (also sometimes called *opt-in sampling*).

*Quota sampling* means that de target population is divided into a number of groups. The population distribution over these groups is supposed to be known. Next, a sample is composed with the same distribution over the groups. The sample sizes in the groups are called *quota*. It is the task of the interviewers to meet these quotas. No random sampling takes place in the groups. It is left to the subjective choice of the interviewers who is interviewed, and who not.

By definition the realized sample is representative with respect to the variables used to form the groups. There is, however, no guarantee at all that the sample will be representative with respect to other variables that were not used to form groups. As a result, estimates of population characteristics may be seriously biased.

Kiaer (1895) was one of the first proposing a form of quota sampling for investigating large populations. He called his approach the *Representative Method*. He divided the population into groups using variables like the degree of urbanization, main industry in the town (agriculture, forestry, industry, seafaring, or fishing) and type of housing (well-to-do houses, middle-class houses, poor-looking houses, or one-person houses). The number of variables was limited because he only could use groups for which the population distribution was known. Kiaer used population distributions that were based on data from a previous census.

As was already mentioned in Chapter 2, the representative method had a major problem: it was impossible to determine the accuracy of estimates. The Polish scientist Jerzy Neyman proved in his famous paper in 1934 that the representative method failed to provide good estimates of population characteristics.

Nevertheless George Gallup kept on using quota sampling for his political polls in the United States in the 1930s and 1940s. He constructed groups based on the variables gender, race, age, and degree of urbanization. All interviewers were assigned quota, like seven white males under 40 living in a rural area, five black males under 40 living in a rural area, six black females under 40 living in a rural area, and so on. Interviewers were free in selecting respondents as long as the quota were met.

The campaign for the presidential election in 1948 proved to be fatal for George Gallup's quota sampling approach. Gallup predicted the Republican candidate Thomas Dewey to be the new president of the United States. This prediction was wrong. The winner of the election was the Democratic candidate Harry Truman. An analysis of the poll showed that Republicans were overrepresented in the sample. Within each group, Republicans were somewhat easier to locate and interview than Democrats. The reason was that Republicans were wealthier than Democrats. Republicans were more likely to have telephones, nicer houses, and permanent addresses. As a result of this incident, George Gallup stopped using quota samples and changed to probability samples.

Quota sampling has the additional problem of hidden nonresponse. Nonresponse occurs if people selected in the sample do not provide the requested information. Usually, respondents differ from nonrespondents. This causes estimates of population characteristics to be biased. To correct for the nasty effects of nonresponse, some kind of correction technique must be carried out. The nonresponse problem is described in more detail in Chapter 7. Nonresponse also occurs in polls based on quota sampling, but it is invisible. Interviewers continue contacting people in the groups until they reach their quota. If people cannot be contacted, refuse to cooperate, or are unable to cooperate, they are simply ignored, and other people are approached.

It can be concluded that quote sampling suffers from a number of shortcomings. These shortcomings prevent generalization of sample results to the target population. Therefore, this type of sampling should be avoided.

## 5.10 SELF-SELECTION

In 1995, it became possible to use the internet for filling in forms, and thus for completing questionnaires. See Bethlehem and Biffignandi (2012, Chapter 1) for more details on the early history. Online polls became quickly popular among polling organizations. This is not surprising as online polls seem to have (at first sight) some advantages

- Now that so many people have internet, an online poll is a simple means to get access to a large group of potential respondents. In some countries almost everybody has internet. For example, internet coverage in The Netherlands and the Scandinavian countries is over 95%.

- An online poll is cheap, because no interviewers are needed. And there are no printing and mailing costs for questionnaires.

- It takes little time to do an online poll. A questionnaire can be launched very quickly. It is sometimes even possible to do an online poll in one day.

- Everybody can do it. There are many software tools on the internet for setting up and carrying out an online poll. Examples are *SurveyMonkey*, *QuestionPro*, *LimeSurvey*, and *Qualtrics*. Some of these tools have even free versions.

So an online poll seems to be an easy, fast, and cheap means for collecting large amounts of data. There are, however, methodological issues. One important issue is sample selection for an online poll. Ideally there is a sampling frame consisting of a list of e-mail addresses of all persons in the target population. A random sample can be selected from the list, and then, an e-mail with a link to the questionnaire can be sent to all selected persons. Unfortunately, such a sampling frame is almost never available. An alternative sample selection procedure could be to select a sample from an address list and to send a letter (by ordinary mail) with a link to the questionnaire to the selected addresses. This makes an online poll more expensive and more time-consuming. With this, some of the advantages of an online poll are lost.

Problems with selecting a random sample for an online poll have caused many researchers to avoid probability sampling. Instead, they rely on *self-selection*. The questionnaire is simply put on the web. Respondents are those people who have internet, happen to visit the website, and spontaneously decide to participate in the poll. So the researcher is not in control of the selection process. Selection probabilities are unknown. Therefore, no unbiased estimates can be computed, nor can the accuracy of estimates be determined.

Self-selection online polls have a high risk of being not representative. There are several reasons. One is that also people from outside the target population can participate. Sometimes it is also possible to complete the questionnaire more than once. It is even possible that certain groups in the population can attempt to manipulate the outcomes of the poll. Here are three examples of online polls with self-selection that were conducted in The Netherlands.

A first example is the election of the 2005 Book of the Year Award (Dutch: NS Publieksprijs), a high-profile literary prize. The winning book was determined by means of a poll on a website. People could vote for one of the nominated books, or mention another book of their choice. About 92,000 people participated in the survey. The winner turned out to be the new interconfessional *Bible* translation launched by The Netherlands and Flanders Bible Societies (see Figure 5.12). This book was not nominated, but nevertheless an overwhelming majority (72%) voted for it. This was due to a campaign launched by (among others) Bible societies, a Christian broadcaster, and Christian newspaper. Although this was all completely within the rules of the contest, the group of voters could hardly be considered representative for the Dutch population.

A second example of a self-selection online poll is an opinion poll during the campaign for the parliamentary elections in The Netherlands in 2012. A group of people tried to influence the outcomes the polls. And by influencing the polls, they hoped to influence the election results. The group consisted of 2500 people. They intended to subscribe to an online opinion panel. Their idea was to behave themselves first as Christian Democrats. Later on, they would change their opinion and vote for the elderly party (50PLUS). They hoped this would affect the opinion of other people too. Unfortunately for them, and fortunately for the researcher,

FIGURE 5.12   The winner of the 2005 Book of the Year Award.

their attempt was discovered when suddenly so many people at the same time subscribed to the panel. See Bronzwaer (2012).

A third example of a self-selection online poll is an opinion poll during the campaign for the local elections in The Netherlands in 2012. A public debate was organized between local party leaders in Amsterdam. A local newspaper, *Het Parool*, conducted an online poll to find out who won the debate. Campaign teams of two parties (the Socialist Party and the Liberal-Democrats) discovered that after disabling cookies it was possible to fill in the questionnaire repeatedly. So, the campaign teams stayed up all night and voted as many times as possible. In the morning, the leaders of these two parties had a disproportionally large number of votes. The newspaper realized that something was wrong and canceled the poll. It accused the two political parties of manipulating the poll. It was the newspaper, however, that had set up a bad poll. See also Bethlehem (2014).

These examples show the dangers of online polls based on self-selection. Therefore, you should be careful when interpreting the result of such polls, or when it is not clear how an online poll was conducted.

## 5.11 SUMMARY

Selecting a sample is a crucial ingredient of a poll. If the sample is selected properly, it is possible to generalize the sample outcomes to the target population. The best way to select such a sample is applying probability sampling. If people are drawn with equal probabilities, the result is a simple random sample.

A simple random sample is representative. The distribution of all variables in the sample will be approximately equal to their population distributions. Be careful to only use the term *representative* if it is explained what it means.

To draw a probability sample, a sampling frame is required. It is a list of all people in the target population. It is important that the sampling frame exactly covers the target population. Unfortunately, sampling frames often suffer from undercoverage, that is, specific parts of the target population are missing. This may affect the outcomes of the poll.

Not every sampling technique is a good one. This chapter pays attention to quota sampling and self-selection sampling. These sampling techniques are used in practice but are not based on probability sampling. Be warned that valid conclusions cannot be drawn about the target population if these sampling techniques were applied.

# CHAPTER 6

# Estimation

## 6.1  ESTIMATOR AND ESTIMATE

The results of a poll are estimates of various population characteristics. Often, only final figures are published and not the way in which they were computed. This raises the question how good these estimates are. Are they close to the true values of the population characteristics? It is important to determine whether or not the estimation procedure had systematic errors that can result in wrong figures. To say it in different words: Was a valid estimation procedure used? Moreover, it should be taken into account that estimates are never exactly equal to the true value to be estimated. Therefore, it should be clear how large the difference at most can be. This is called the *margin of error*.

The documentation of a poll should provide sufficient information to give an indication of the accuracy of the estimates. Preferably, the poll report should include margins of errors. Unfortunately, this is not always the case. This chapter gives some background on estimation procedures. It will also be shown how the accuracy of the estimates can be computed.

The first step in the analysis of the poll data will usually be the estimation of *population characteristics*, such as totals, means, and percentages of target variables. It may be informative to compute these estimates also for several subpopulations into which the target population can be divided. For example, separate estimates can be computed for males and females, for various age groups, or for various regions in the country. One should always realize that only estimates of population characteristics can be computed, and not their exact values. This is because only data about a sample of people from the target population are available. If a probability sample was drawn, valid estimates can be computed, and also, the accuracy of the estimates can be computed (e.g., in the form of margins of error). It is important that the researcher publishes these margins of error, if only to avoid the impression that his estimates present true values.

To compute an estimate, an estimator is required. An *estimator* is a procedure, a recipe, describing how to compute an *estimate*. The recipe also makes clear which ingredients are required. Of course, the researcher must have the sample values of the target variables. Sometimes, he has additional information, such as sample values and the population distribution of auxiliary variables.

An estimator is only meaningful if it produces estimates that are close to the population characteristics to be estimated. Therefore, a good estimator must satisfy the following two conditions:

1. An estimator must be *unbiased*. Suppose, the process of drawing a sample and computing an estimate is repeated a large number of times. As random samples are drawn, each turn will result in a different sample. Therefore, the estimate will also be different. The estimator is unbiased if the average of all the estimates is equal to the population characteristic to be estimated. To say it in different words: The estimator must not systematically over- or underestimate the true population value.

2. An estimator must be *precise*. All estimates obtained by repeatedly selecting a sample must be close to each other. In other words: The variation in the outcomes of the estimator must be as small as possible.

The term *valid* is related to the term *unbiased*. A measuring instrument is called valid if it measures what it intends to measure. Hence, an unbiased estimator is a valid measuring instrument, because it estimates what it intends to estimate.

The term *reliable* is related to the term *precision*. A measuring instrument is called reliable if repeated use leads to (approximately) the same value. Hence, a precise estimator is a reliable measurement instrument, because all possible estimates are close to each other.

## 6.2 AN EXAMPLE OF A POLL

It can be shown theoretically that probability sampling works. This requires application of some mathematics and probability theory. It can also be shown in a different way, by simulating the process of sample selection and computing estimates. This is what is done in this section.

Elections will be held in the town of Rhinewood. The number of eligible voters is 30,000. There is a new political party in Rhinewood called the New Internet Party (NIP). This party aims at modernizing democratic processes in the town by using the internet. Not surprisingly, the party expects many votes from internet users.

To be able to simulate sampling, a fictitious population was created consisting of 30,000 people. The population was constructed in such a way that 39.5% of these people vote for the NIP. Of course, this percentage is unknown in practice. The question is now how well it can be estimated by drawing a random sample.

First, random samples of size 500 are selected. For each sample, the percentage of voters for the NIP is computed. The results of 900 samples are used to show what the possible outcomes of the estimator can be. The results are presented in the form of a histogram. Each little block represents one estimate. If estimates are very close, they are stacked onto each other. The graph in Figure 6.1a shows the distribution of the 900 estimates. The vertical black line at 39.5 represents the population value to be estimated.

It is clear that the estimates are concentrated around the value to be estimated. Some estimates are smaller than the true value, and other estimates are larger. On average, the estimates are equal to the population percentage of 39.5%. So, the estimator is unbiased. There is still variation in the estimates. The values are spread between approximately 34% and 46%. So, the precision is not very high.

What happens if the sample size is increased from 500 to 2000? Again, 900 samples are drawn, and the estimate is computed for each sample. The graph in Figure 6.1b shows the results. The estimator remains unbiased, but the values are now much more concentrated around the true population value. So, the estimator is much more precise. The results in Figure 6.1 reflect an important principle: Estimates are more precise if the sample size is increased.

What happens if the poll is an online poll? Suppose, samples are selected from only the group of people who have access to the internet. The graph

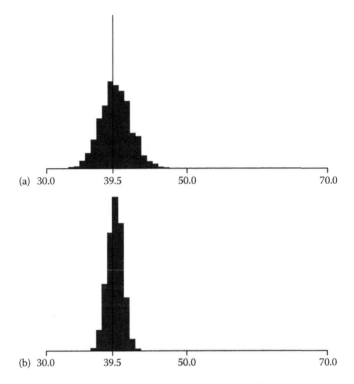

FIGURE 6.1    Estimating a percentage with simple random samples of size 500 (a) and 2000 (b).

in Figure 6.2a shows the results for 900 samples of size 500. It turns out that all values are systematically too high. Their average value is 56.4%. So, the estimator has a bias of 56.4% − 39.5% = 16.9%.

This bias is not surprising. All people in the sample have internet, and they are typically people that tend to vote for the NIP. So, NIP-voters are overrepresented in the samples. The values of the estimates vary between 51% and 63%, which is approximately the same amount of variation as in the graph of Figure 6.1a.

The graph in Figure 6.2b shows what happens if the sample size of the online poll is increased from 500 to 2000. There is clearly less variation, so the estimator is more precise. Again, increasing the sample size improves the precision of the estimator. Note that increasing the sample size did not decrease the bias. The estimates are, on average, still 16.9%, which is too high. The important message is that increasing the sample size does not help to reduce an existing bias.

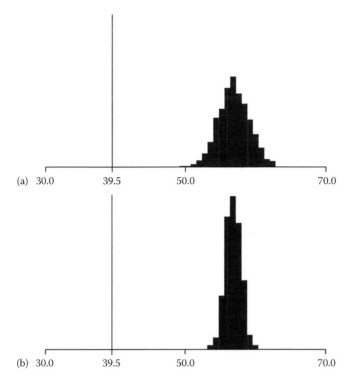

FIGURE 6.2   Estimating a percentage with an online poll. Samples of size 500 (a) and 2000 (b).

## 6.3 ESTIMATING A POPULATION PERCENTAGE

A percentage is a population characteristic that is estimated very often in a poll. Examples are the percentage of voters for a specific party, the percentage of people in the United Kingdom wanting the United Kingdom to leave the European Union, and the percentage of people in favor of a certain government measure. If the sample is a probability sample, and people are selected with equal probabilities, the *analogy principle* applies. This principle means that a population characteristic can be estimated by computing the corresponding sample characteristic. So, the sample percentage is a good estimate for the population percentage. For example, if a researcher wants to estimate the percentage of people in the country having trust in the European Union, he should use the corresponding percentage in the sample as an estimate.

The sample percentage is (under simple random sampling without replacement) an unbiased estimator of the population percentage. This can be proved by applying some mathematics and probability theory. There is also another way to show it, and this is by carrying out a simulation: construct a population, select many samples, compute the estimate for each sample, and see how the estimates behave. Section 6.2 contains an example of such a simulation. It was shown there that the percentage of voters for the NIP in the sample is an unbiased estimator for the percentage of NIP voters in the population.

Once an estimate for a population characteristic is obtained, the margin of error of this estimate must also be computed. A basic ingredient of the margin of error is the *variance of the estimator*. The mathematical expression of the variance of the sample percentage is equal to

$$\text{Variance} = \left(\frac{1}{n} - \frac{1}{N}\right) \times \frac{N}{N-1} \times P \times (100 - P)$$

where:
P is the percentage in the population
N is the size of the population
n is the sample size

P is the population characteristic to be estimated. Therefore, its value is unknown. Consequently, it is not possible to compute the variance. The way out is replacing the variance by its sample-based estimated variance. The estimated variance can be calculated by the following equation:

$$\text{Estimated variance} = \left(\frac{1}{n} - \frac{1}{N}\right) \times \frac{n}{n-1} \times p \times (100 - p)$$

Note that the population percentage P has been replaced by the sample percentage p, and the population size N has been replaced by the sample size n. If the population is very large and the sample is much smaller than the population, the following simple approximation of the estimated variance can be used:

$$\text{Estimated variance} \approx \frac{p \times (100 - p)}{n-1}$$

Once one knows the estimated variance, the estimated standard error can be computed by taking the square root:

$$\text{Estimated standard error} = \sqrt{\text{Estimated variance}}$$

Finally, the margin of error is obtained by multiplying the estimated standard error by a factor 1.96. This results in

$$\text{Margin of error} = 1.96 \times \text{Estimated standard error}$$

If the approximation for the estimated variance is used and the factor 1.96 is rounded to 2, the result is a simple formula for the margin of error:

$$\text{Margin of error} \approx 2 \times \sqrt{\frac{p \times (100 - p)}{n - 1}}$$

The margin of error indicates how large the difference can at most be between the estimate and the true, but unknown, population value. Here is an example. Suppose, a sample of size $n = 500$ is drawn from the population of voters in Rhinewood. It turns out that 200 people in the sample say they will vote for the NIP. Hence, the sample percentage $p$ is equal to 40%. Then the margin of error is equal to

$$\text{Margin of error} \approx 2 \times \sqrt{\frac{40 \times (100 - 40)}{500 - 1}} = 2 \times \sqrt{\frac{2400}{499}} = 2 \times \sqrt{4.81} = 4.4$$

A margin of error of 4.4 means that the percentage of NIP voters in the population cannot differ more than 4.4 from its estimate 40%. To say it in other words, the percentage of voters for the NIP in the population is somewhere between 35.6% and 44.4%. This interval is called the *confidence interval*. Its lower bound is obtained by subtracting the margin of error from the estimate and the upper bound by adding the margin of error to the estimate.

It should be said that statements about the margin of error or the confidence interval always have an element of uncertainty. The correct statement is that the confidence interval contains the true population value with a high probability. This probability is here equal to 95%. Therefore, it is better to call this confidence interval the 95% *confidence interval*.

Use of a 95% confidence interval means that the confidence interval contains the true population value with a probability of 95%. This implies that in 5% of the cases, a wrong conclusion is drawn from the confidence interval (the true value lies in the interval while this is not the case). If a higher confidence level is required, one can, for example, compute a 99% confidence interval. This interval is obtained by replacing the value 1.96 in the above-mentioned formulae by the value 2.58.

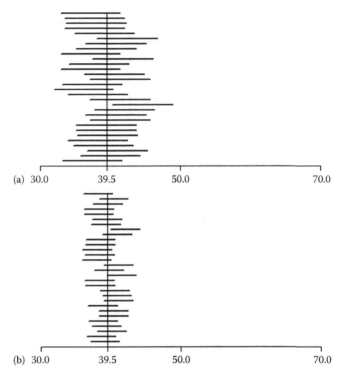

FIGURE 6.3   Confidence intervals for a percentage for samples of size 500 (a) and 2000 (b).

Figure 6.3 shows the result of simulating confidence intervals for an opinion poll in the fictitious town of Rhinewood. The target population consists of 30,000 voters. Objective of the poll is estimation of the percentage of voters for the NIP. The graph in Figure 6.3a shows the confidence intervals for 30 samples of size 500. The vertical black line denotes the population percentage (39.5%) to be estimated. The horizontal line segments represent the 95% confidence intervals. Almost all intervals contain the population percentage. Only 1 interval out of 30 (3.3%) makes the wrong prediction.

The graph in Figure 6.3b shows the 95% confidence intervals for 30 samples of size 2000. It is clear that these confidence intervals are much smaller than those for samples of size 500. This proves again that increasing the sample size improves the precision of estimators. Again, there is one confidence interval, which is not containing the true population percentage. This reflects the confidence level of 95%. On average, 1 out of 20 times a wrong conclusion will be drawn from the confidence interval.

TABLE 6.1 Computation of the Margin of Error and the Confidence Interval

| Step | Ingredients:<br>n: Sample Size<br>c: Number of People in the Sample<br>with a Specific Property | Formula | Example:<br>n = 500<br>c = 200 |
|---|---|---|---|
| 1 | Compute sample percentage | $p = 100 \times c/n$ | $p = 100 \times 200/500 = 40$ |
| 2 | Compute complementary percentage | $q = 100 - p$ | $q = 100 - 40 = 60$ |
| 3 | Multiply both percentages | $r = p \times q$ | $r = 40 \times 60 = 2400$ |
| 4 | Compute the estimated variance by dividing the result by the sample size minus 1 | $v = r/(n-1)$ | $v = 2400/499 = 4.8$ |
| 5 | Compute the estimated standard error by taking the square root of the result | $s = \sqrt{v}$ | $s = \sqrt{4.8} = 2.2$ |
| 6 | Compute the margin of error by multiplying the result by 2 | $m = 2 \times s$ | $m = 2 \times 2.2 = 4.4$ |
| 7 | Compute the lower bound of the confidence interval by subtracting the margin of error from the estimate | $lb = p - m$ | $lb = 40 - 4.4 = 35.6$ |
| 8 | Compute the upper bound of the confidence interval by adding the margin of error to the estimate | $ub = p + m$ | $ub = 40 + 4.4 = 44.4$ |

The computations for the 95% confidence interval for a population percentage are summarized in Table 6.1. The rightmost column (column 4) contains an example. Note that the simple approximations for the margin of error and the confidence interval were applied here. This is allowed because the population size (30,000) is much larger than the sample size (500).

## 6.4 ESTIMATING A POPULATION MEAN

Another type of population characteristic, which is often estimated, is the mean of a *quantitative variable*. Such a variable is usually measured with a numerical question. Examples are the mean amount of time spent per day on the internet, the mean monthly income, and the mean household size. The *population mean* of a variable is obtained by adding all values and dividing the sum by the size of the population.

How to estimate the population mean? If a simple random sample was drawn, the analogy principle applies: The sample mean is a good estimator for the population mean. For example, if a researcher wants to estimate how many hours per week people listen on average to a local radio station, he can use the mean number of hours listened in the sample for this.

The sample mean is an unbiased estimator of the population mean (under simple random sampling without replacement). This can be proved mathematically, but it can also be shown by carrying out a simulation: construct a fictitious population, select many samples, compute the estimate for each sample, and see how the estimates behave.

A radio-listening poll was simulated. A population of 15,000 people in the fictitious town of Harewood was created. In this town, 8535 people listened to the local radio station. For each of the 8535 listeners, the population file contained the number of hours listened last week. For the nonlisteners, the number of hours was set to 0. A large number of samples was selected (with simple random sampling), and the mean number of hours listened was computed in each sample. Thus, a large number of estimates were obtained. The values of these estimates were displayed in the form of a histogram. The graph in Figure 6.4a shows the result of drawing 1000 samples of size 500. The vertical black line represents the population value to be estimated: 2.7 hours. The estimates are nicely distributed around this value. Many estimates are close to it.

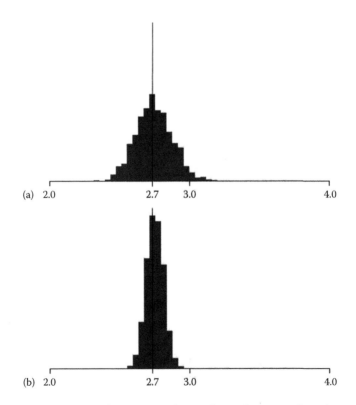

FIGURE 6.4 Estimating the mean with simple random samples of size 500 (a) and 2000 (b).

On average, the estimates are equal to this value. Therefore, the sample mean is an unbiased estimator.

What happens if the sample size is increased? The graph in Figure 6.4b shows what happens if the sample size is 2000. The estimates are closer to the true population value. There is no systematic under- or overestimation. The conclusion is that the sample means still is an unbiased estimator of the population mean. Moreover, this estimator is more precise because the sample size is larger.

Once an estimate for the population mean has been obtained, its margin of error must be computed. A basic ingredient of the margin of error is the variance of the estimator. The mathematical expression of the variance of the sample mean is as follows:

$$\text{Variance} = \left(\frac{1}{n} - \frac{1}{N}\right) \times S^2$$

where:

$$S^2 = \frac{(Y_1 - \text{Pmean})^2 + (Y_2 - \text{Pmean})^2 + \ldots + (Y_N - \text{Pmean})^2}{N-1}$$

The size of the population is denoted by $N$, and the size of the sample by $n$. $Y_1, Y_2, \ldots, Y_N$ are all $N$ values of the variable in the population, and Pmean is the population mean of all these values.

The quantity $S^2$ cannot be computed in practice, because not all values of the variable are known. Only the values in the sample are available. If all values were known, the poll would not have been necessary. Hence, it is not possible to compute the variance. The way out is replacing the variance by its sample-based estimator. The estimated variance can be calculated as follows:

$$\text{Estimated variance} = \left(\frac{1}{n} - \frac{1}{N}\right) \times s^2$$

where:

$$s^2 = \frac{(y_1 - \text{Smean})^2 + (y_2 - \text{Smean})^2 + \ldots + (y_n - \text{Smean})^2}{n-1}$$

The quantities $y_1, y_2, \ldots, y_n$ denote the $n$ values of the variable in the sample, and Smean is the mean of these values. If the population is very large, and the sample is much smaller than the population, the following simple approximation of the estimated variance can be used:

$$\text{Estimated variance} = \frac{s^2}{n}$$

Once the estimated variance is obtained, the estimated standard error is computed by taking the square root:

$$\text{Estimated standard error} = \sqrt{\text{Estimated variance}}$$

Finally, the margin of error is determined by multiplying the estimated standard error by the factor 1.96. This results in

$$\text{Margin of error} = 1.96 \times \text{Estimated standard error}$$

If the approximation for the estimated variance is used, the factor 1.96 is replaced by 2, a simple formula for the margin of error is obtained:

$$\text{Margin of error} \approx 2 \times \sqrt{\frac{s^2}{n}}$$

The above-mentioned expressions describe how to compute a 95% confidence interval. This implies that a wrong conclusion will be drawn in 5% of the cases. If the risk of drawing a wrong conclusion is reduced, a 99% confidence interval can be used. The only thing that must be changed is replacing the factor 1.96 in the expression for the margin of error by the factor 2.58.

Here is an example of how to compute a confidence interval manually. Suppose, a simple random sample of 20 people was drawn from the target population of 15,000 people in the town of Harewood. Objective of the poll was exploring listening behavior of its inhabitants. The people in the poll were asked how many hours they listened to the local radio station last week. The answers are in the second column (*Value*) of Table 6.2.

The sample mean is equal to the sum of the values in the second column (56.40) divided by 20. The result is 2.82 (hours).

To estimate the variance of the estimate, first, the sample variance $s^2$ must be computed. To do this, the sample mean (2.82) is subtracted from each value. This leads to column 3. Next, take the square of each value in column 3. The result can be found in column 4. Now, the sample variance

TABLE 6.2  A Sample of 20 Persons from the Population
of Harewood

| No. | Value | Value – Smean | (Value – Smean)$^2$ |
|-----|-------|---------------|---------------------|
| 1 | 0.00 | −2.82 | 7.95 |
| 2 | 3.40 | 0.58 | 0.34 |
| 3 | 4.60 | 1.78 | 3.17 |
| 4 | 4.10 | 1.28 | 1.64 |
| 5 | 3.90 | 1.08 | 1.17 |
| 6 | 0.00 | −2.82 | 7.95 |
| 7 | 3.40 | 0.58 | 0.34 |
| 8 | 7.30 | 4.48 | 20.07 |
| 9 | 0.00 | −2.82 | 7.95 |
| 10 | 0.00 | −2.82 | 7.95 |
| 11 | 3.80 | 0.98 | 0.96 |
| 12 | 0.00 | −2.82 | 7.95 |
| 13 | 0.00 | −2.82 | 7.95 |
| 14 | 4.20 | 1.38 | 1.90 |
| 15 | 5.50 | 2.68 | 7.18 |
| 16 | 4.40 | 1.58 | 2.50 |
| 17 | 6.40 | 3.58 | 12.82 |
| 18 | 5.40 | 2.58 | 6.66 |
| 19 | 0.00 | −2.82 | 7.95 |
| 20 | 0.00 | −2.82 | 7.95 |
| Sum | 56.40 | 0.00 | 122.35 |

is obtained by summing the values in column 4 and dividing the sum by $n − 1 = 19$. This gives $s^2 = 122.35/19 = 6.44$.

The estimated variance of the estimator is obtained by multiplying $s^2$ by $(1/n − 1/N) = (1/20 − 1/15,000)$. This results in the value 0.32. The estimated standard error is now equal to the square root of 0.32, and this is 0.57.

The margin of error of the 95% confidence interval is equal to the standard error multiplied by 1.96. The (rounded) result is 1.11. The lower bound of the 95% confidence interval is equal to $2.82 − 1.11 = 1.71$. The upper bound is equal to $2.82 + 1.11 = 3.93$.

The conclusion is that with a high probability (95%), the mean number of hours listened to the local radio station last week is somewhere between 1.71 and 3.93 hours. The estimate has a substantial margin of error of 1.11 hours. This is caused by the small sample size of 20 people. The margin of error can be decreased by taking a larger sample.

Manual computation of estimates and margins of errors is a cumbersome procedure for larger polls. Fortunately, there are all kinds of software packages that can do it. One example is the function *STDEV.S* in the spreadsheet package MS Excel. Of course, there are also many statistical packages, such as SPSS, Stata, SAS, and R that can do it.

## 6.5 HOW LARGE SHOULD THE SAMPLE BE?

A researcher has to determine in the design phase of the poll how large the size of the sample must be. This is an important decision. If, on the one hand, the sample is larger than what is really necessary, a lot of time and money may be wasted. And if, on the other, the sample is too small, estimates are not as precise as planned, and this makes the results of the poll less useful. Considering the possible consequences of the choice of the sample size, it should always be mentioned in the poll documentation.

It is not so simple to determine the sample size as it depends on a number of different factors. It was already shown that there is a relationship between the precision of estimators and the sample size: the larger the sample, the more precise the estimators. Therefore, the question about the sample size can only be answered if it is clear how precise the estimators must be. Once the precision has been specified, the sample size can be computed. A very high precision nearly always requires a large sample. However, a large poll will also be costly and time consuming. Therefore, the sample size will in many practical situations be a compromise between costs and precision.

This section explains how to compute the sample size for a simple random without replacement. First, the case of estimating a population percentage is considered. Next, the case of estimating a population mean is described.

### 6.5.1 The Sample Size for Estimating a Percentage

Starting point is the assumption that there is some indication of how large the *margin of error* at most can be. The margin of error is defined as the distance between the estimate and the lower or upper bound of the confidence interval. So, the margin of error is the maximum allowed difference between estimate and true value.

The margin of error of a 95% confidence interval is defined as the standard error of the estimator multiplied by 1.96. Therefore, setting a maximum to the margin of error implies that the standard error may not exceed a certain value. Suppose, Max is the value of the margin of error that may not be exceeded. This implies that the following inequality must hold:

$$1.96 \times \text{Standard error} \leq \text{Max}$$

The standard error of a percentage in the sample can be calculated using the following equation:

$$\text{Standard error} = \left(\frac{1}{n} - \frac{1}{N}\right) \times \frac{N}{N-1} \times P \times (100 - P)$$

Transforming this result into an expression for $n$ leads to

$$n \geq \frac{1}{\dfrac{N-1}{N} \times \left(\dfrac{\text{Max}}{1.96}\right)^2 \times \dfrac{1}{P(100-P)} + \dfrac{1}{N}}$$

If the size $N$ of the target population is large, this expression can be simplified to

$$n \geq \left(\frac{1.96}{\text{Max}}\right)^2 \times P \times (100 - P)$$

A problem of both expressions is that they contain the unknown population percentage $P$. Indeed, getting to know this value was the reason to conduct the poll. To be able to use one of the expressions for computing the sample size, some indication of the value of $P$ must be obtained and substituted. This could, for example, be a value from a previous poll. If there is no idea at all of the value of the population percentage, the value $P = 50$ can be substituted. This value leads to the largest sample size. If the maximum margin of error is not exceeded for $P = 50$, it also will not happen for any other value of $P$.

Let us return to the town of Rhinewood, where elections will be held. The number of eligible voters is 30,000. There is a new political party in Rhinewood called the NIP. This party aims at modernizing democratic processes in the town using the internet. Not surprisingly, the party expects many votes from internet users. A previous poll predicted that approximately 30% would vote for the NIP.

A new election poll is carried out to explore the popularity of the new party. The margin of error should not exceed 3%. If the expression for the sample size is applied with $N = 30,000$, $P = 30$, and Max = 3, the result is

$$n \geq \frac{1}{(29,999/30,000) \times (3/1.96)^2 \times (1/30 \times 70) + (1/30,000)} = 870.4$$

Rounding up this value to the nearest integer number gives a minimal sample size of 871. Application of the simplified expression would lead to a minimal sample size of 897.

If nothing is known about the value of $P$, use the value $P = 50$. The expression for the sample size now becomes

$$n \geq \frac{1}{(29,999/30,000) \times (3/1.96)^2 \times [1/50(50)] + (1/30,000)} = 1030.5$$

Rounding up this value to the nearest integer number gives a minimal sample size of 1031. The simplified formula leads to a minimal sample size of 1068.

Instead of computing the minimal sample size with one of the expressions, Table 6.3 can also be used. The required sample size can be read from this table, given an indication of the population percentage to be estimated and the maximal margin of error.

Many polls work with sample sizes of around 1000 people. Table 6.3 shows that for such a sample size, the margin of error will not exceed 3%.

TABLE 6.3    Required Sample Size for Estimating a Percentage

| Indication of the Population Percentage | Maximal Margin of Error | | | | |
|---|---|---|---|---|---|
| | 1 | 2 | 3 | 4 | 5 |
| 5 | 1825 | 457 | 203 | 115 | 73 |
| 10 | 3458 | 865 | 385 | 217 | 139 |
| 15 | 4899 | 1225 | 545 | 307 | 196 |
| 20 | 6147 | 1537 | 683 | 385 | 246 |
| 25 | 7204 | 1801 | 801 | 451 | 289 |
| 30 | 8068 | 2017 | 897 | 505 | 323 |
| 35 | 8740 | 2185 | 972 | 547 | 350 |
| 40 | 9220 | 2305 | 1025 | 577 | 369 |
| 45 | 9508 | 2377 | 1057 | 595 | 381 |
| 50 | 9605 | 2402 | **1068** | 601 | 385 |
| 55 | 9508 | 2377 | 1057 | 595 | 381 |
| 60 | 9220 | 2305 | 1025 | 577 | 369 |
| 65 | 8740 | 2185 | 972 | 547 | 350 |
| 70 | 8068 | 2017 | 897 | 505 | 323 |
| 75 | 7204 | 1801 | 801 | 451 | 289 |
| 80 | 6147 | 1537 | 683 | 385 | 246 |
| 85 | 4899 | 1225 | 545 | 307 | 196 |
| 90 | 3458 | 865 | 385 | 217 | 139 |
| 95 | 1825 | 457 | 203 | 115 | 73 |

If the sample size is 1068 as indicated in the table, the margin of error for every percentage will at most be 3%; or to say it otherwise: With a sample of 1068 persons, the estimate will never differ more than 3% from the true population value. So, if the sample percentage is, for example, equal to 48%, the population percentage will be between 45% and 51% (with a high probability of 95%).

## 6.5.2 The Sample Size for Estimating a Mean

For estimating the population mean, the starting point is also the assumption that some indication is available of how large the *margin of error* at most may be. The margin of error is defined as the distance between the estimate and the lower or upper bound of the confidence interval. So, the margin of error is the maximum allowed difference between estimate and true value.

The margin of error of a 95% confidence interval is defined as the estimated standard error of the estimator multiplied by 1.96. Therefore, setting a maximum to the margin of error implies that the estimated standard error may not exceed a certain value.

Suppose, Max is the value of the margin of error that may not be exceeded. This implies that

$$1.96 \times \text{Estimated standard error} \leq \text{Max}$$

Transforming this result into an expression for $n$ leads to

$$n \geq \frac{1}{(\text{Max}/1.96)^2 \times (1/S^2) + (1/N)}$$

in which the quantity $S^2$ is defined by

$$S^2 = \frac{(Y_1 - \text{Pmean})^2 + (Y_2 - \text{Pmean})^2 + \ldots + (Y_N - \text{Pmean})^2}{N-1}$$

See also Section 6.4. If the size $N$ of the target population is large, this expression can be simplified to

$$n \geq \left(\frac{1.96}{\text{Max}}\right)^2 \times S^2$$

Both expressions contain the quantity $S^2$. In practice, this value is not known. This makes it difficult to apply these expressions. Sometimes, the researcher may have some indication of its value from a different survey, or from a small test survey.

If it is known that the values of the target variable have a more or less a normal distribution (a bell-shaped, symmetric distribution) over an interval of length $L$, the quantity $(L/6)^2$ as a rough approximation for $S^2$.

## 6.6 SUMMARY

A poll is conducted to increase knowledge about a specific population. To achieve this, people in the population are asked some specific questions. Together, all these questions comprise the questionnaire. It costs a lot of time and money to let everybody fill in a questionnaire. Therefore, a sample is selected from the population, and only the people in the sample complete the questionnaire.

The sample data must be used to say something about the population as a whole. This means generalizing from the sample to the population. Generalization usually takes the form of estimating various population characteristics. An estimator is a recipe for computing such estimates.

A good estimator is unbiased. This means that the estimator produces on average the true population value. There is no systematic under- or overestimation. A good estimator is also precise. This means that all possible values of the estimator are close to each other.

Good estimators can be obtained if the sample is selected by means of probability sampling. In the case of simple random sampling with equal probabilities, the analogy principle applies: A population characteristic can be estimated by computing the corresponding sample characteristic. So, the sample percentage is a good estimator for the population percentage, and the sample mean is a good estimator for the population mean.

An estimate is almost never exactly equal to the true population value. There is always an element of uncertainty. This uncertainty can be quantified in the form of the margin of error. The margin of error indicates how far away the estimate can at most be from the true population value. The margin of error can also be used to compute a confidence interval. This confidence interval contains with a high probability the true population value.

If a researcher intends to carry out a poll, he must decide how large the sample will be. The sample size depends on how precise he wants the estimates to be. If the sample size increases, estimates will be more precise. A rule of thumb is that for a sample of 1000 people, the margin of error of an estimate of a percentage is around 3%.

# CHAPTER 7

# Nonresponse

## 7.1 THE NONRESPONSE PROBLEM

The previous chapters explained what makes a poll a good poll. The questionnaire is one important ingredient. Without a good questionnaire, researchers run a serious risk of getting wrong answers to their questions. Consequently, wrong conclusions will be drawn from the poll data. Another important aspect of a poll is the way in which the sample is selected. It will be clear by now that the best sample is a probability sample. And if all people have the same selection probability, the analogy principle can be applied so that simple estimates can be computed.

Polls based on a good questionnaire and simple random sampling make it possible to compute valid estimates of population characteristics. And moreover, it is also possible to compute the margins of error, which is an indication of the precision of the estimates. The poll will produce valid and reliable results that can be generalized to the target population. Unfortunately, life is not so simple. There are always practical problems that affect the outcomes of polls. One of the most important problems is nonresponse. *Nonresponse* occurs when people in the selected sample do not provide the requested information, or when the provided information is unusable.

Nonresponse may have a serious impact on the outcomes of the poll. Therefore, it is important to prevent it from happening as much as possible. Unfortunately, whatever the efforts, there will always be nonresponse. And if it is not possible to prevent nonresponse, something has to be done to remove or reduce a possible bias of the estimates of population characteristics.

Taking into account the possible impact on the outcomes of a poll, researchers should always give information about nonresponse in their poll reports. They should at least report the response rate. And if there is a serious amount of nonresponse, they should carry out an analysis of the nonresponse. This analysis should make clear whether the outcomes of the poll may be affected.

If a poll is affected by nonresponse, the researcher should attempt to repair these problems. The most popular way to do this is carrying out some form of weighting adjustment. Of course, the poll report should document this correction procedure.

Sometimes, one encounters poll reports not mentioning nonresponse. They seem to suggest that there was no nonresponse, and therefore, there were no nonresponse problems. This is highly unlikely. Almost every poll is affected by nonresponse. So be careful when a poll report does not report the response rate.

An example is the Eurobarometer. This is a series of opinion polls commissioned by the European Commission. The standard Eurobarometer poll is conducted twice a year. It is a face-to-face poll, in which interviewers visit randomly selected addresses. The planned sample size is 1000 respondents in each country (with the exception of a few small countries). Surprisingly, the realized sample size is also 1000 respondents per country. This would imply that people at the selected addresses are always home, always agree to cooperate, are never ill, and have never language problems. Of course, this is not the case. Although not documented, it is more likely that interviewers continued their interview attempts until they reached their quotas. And in the process, they must have faced a lot of nonresponse.

Another example is the polls in the campaign for the general election in the United Kingdom on May 7, 2015. Some of these polls were telephone polls. The samples for these polls were selected by means of random digit dialing (RDD). Response rates were not reported, but Martin Boon of the polling organization ICM mentioned on television that they had to call 30,000 random numbers in order to get 2000 respondents. See Reuben (2015). This comes down to a response rate of only 7%. Not surprisingly, all telephone polls failed to correctly predict the outcome of the election.

This chapter gives some more background about the nonresponse problem. It shows what the possible effects of nonresponse can be. It also makes clear that it is important to take a look at nonresponse in polls. And this chapter contains a simple introduction to weighting adjustment techniques. The chapter is only about unit nonresponse and not about item nonresponse. *Unit nonresponse* occurs when a selected person does not provide any information at all. The questionnaire remains completely empty. Another type of nonresponse is *item nonresponse.* This occurs when some questions have been answered, but no answer is obtained for some other, possibly sensitive, questions.

## 7.2 CONSEQUENCES OF NONRESPONSE

One obvious consequence of nonresponse is that the realized sample is smaller than planned. If the researcher wants to have a sample of size 1000, and he draws 1000 people from the sampling frame, but half of them respond, the realized sample size will be only 500. A smaller sample size will decrease the precision of an estimator. But valid estimates can still be obtained as the computed margins of errors or confidence intervals take into account the smaller sample size.

The problem if it be a too small sample can easily be solved. Just start with a larger initial sample size. For example, if a response of 1000 people is required in a poll, and the expected response rate is around 50%, the initial sample size should be approximately 2000 people.

A far more serious consequence of nonresponse is that estimates of population characteristics can be biased. This situation occurs if, due to nonresponse, some groups in the population are under- or overrepresented in the sample, and these groups behave differently with respect to the phenomena being investigated. This is called *selective nonresponse.*

If there is nonresponse in a poll, it is likely that estimates are biased unless there is very convincing evidence to the contrary. Here are a few examples:

- Data collection in a victimization survey was carried out by means of face-to-face interviews. It turned out that people, who are afraid to be at home alone at night, were less inclined to participate in the survey. They simply did not open the door when the interviewer called at night.

- People, who refused to participate in a housing demand survey, had lesser housing demands than those who responded.

- Mobile people were underrepresented in a mobility survey. They more often were not at home when the interviewer called.

- In many election polls, voters are overrepresented among the respondents, and nonvoters are overrepresented among the nonrespondents.

A simulation shows what the effects of nonresponse can be. The fictitious population of 30,000 voters in Rhinewood is used in this experiment. The population was constructed in such a way that 39.5% of the people vote for the New Internet Party (NIP). First, random samples of size 500 were drawn in which all selected people responded (the case of full response). For each sample, the percentage of voters for the NIP was computed. The estimates of 900 samples were used to show what the possible outcomes of the estimator can be. The results are shown in the graph of Figure 7.1a. The estimates are nicely distributed around the value to be estimated (indicated by the vertical black line). Many estimates are close to this value. On average, the estimates are equal to this value. Therefore, it can be concluded that the sample percentage is an unbiased estimator.

Now, nonresponse is generated. This is done in such a way that response is high among people with internet and low among those without it. Again,

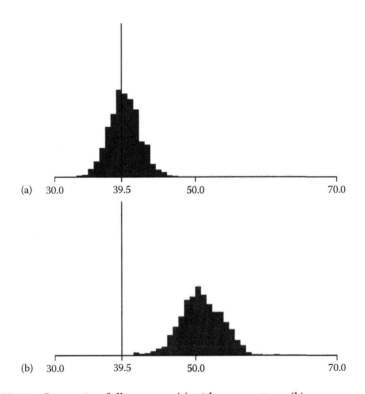

FIGURE 7.1   Comparing full response (a) with nonresponse (b).

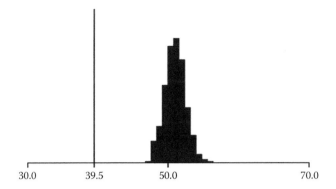

FIGURE 7.2   The effect of the sample size on nonresponse bias.

900 samples of size 500 were drawn, and percentages of NIP-voters were computed. The distribution of these estimates is shown in the graph of Figure 7.1b.

The distribution is clearly not concentrated around the true population value (39.5%). The distribution has shifted to the right. The average estimate is now 50.6%. So there is a bias of 50.6% − 39.5% = 11.1%. The cause of this bias is that people with internet tend to vote for the NIP, and these people are overrepresented in the poll.

Note that the variation of the estimates is larger in the graph of Figure 7.1b. This is due to the smaller sample size. The response rate was approximately 58%, which corresponds to a sample size of 290.

Sometimes, people think that nonresponse problems can be solved by increasing the sample size. This is not the case, and this shown in the graph in Figure 7.2. This is the result of a simulation of nonresponse similar to the one in Figure 7.1, but now, the sample size has been increased from 500 to 2000. The bias remains the same, but the precision increased. So, one could say that the estimators more precisely estimate the wrong value.

The problems of nonresponse have not diminished over time. To the contrary, surveys and polls in many countries suffer from increased nonresponse rates. As an example, Figure 7.3 shows the response rate over the years of one of the most important survey of Statistics Netherlands. It is the Labor Force Survey (LFS). The LFS is conducted in all member states of the European Union. It collects information on labor participation of people aged 15 and over, as well as on persons outside the labor force. The response was high in the 1970s (almost 90%), but it declined over the years. Nowadays, the response rate is below 60%.

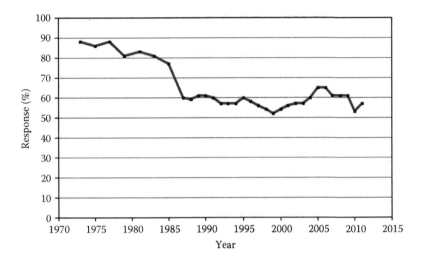

FIGURE 7.3   The response rate in the Dutch Labor Force Survey.

Nowadays, it is not easy to obtain a response rate of 60%. It requires con-
siderable efforts.

The response rate of a poll depends on many factors. Some of them are
described here. There are probably a lot more factors influencing response
behavior.

An important factor is the topic of the poll. If people are interested in
the topic of the poll, they are more inclined to participate. The chances of
success are much smaller for dull, uninteresting, or irrelevant polls. For a
more detailed treatment for the effects of the topic of a poll, see, for example, Groves et al. (2004).

Who is behind the poll? If a poll is commissioned by a government
organization, the response rate often tends to be higher than those of polls
of commercial market research companies. One of the reasons probably is
that people consider government polls more important and more relevant.

If the target population of a poll consists of households, questions can
be answered by any person in the household (from a certain age). This
makes it easier to collect information about the household, and thus, this
increases the probability of response. If a specific person in the household
is required to complete the questionnaire, establishing contact and getting
cooperation can be much harder.

Response rates differ by country. An example is the European Social
Survey (ESS) that is regularly conducted in many European countries. In each

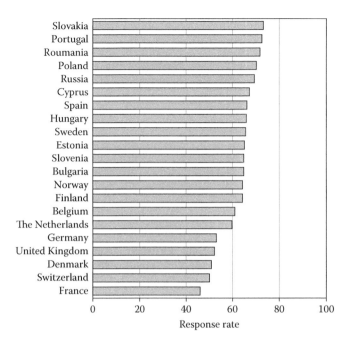

FIGURE 7.4 Response rates in the third round of the European Social Survey. (From Stoop et al., 2010.)

country, the design of the survey is similar. Stoop et al. (2010) provide the response rates for the third round of the ESS. They are shown in Figure 7.4. There are clear differences. For example, the response rate is over 70% in Slovakia, and in France, it is under 50%.

Even the time of the year has an impact on the response rate. It is wise to avoid the summer holidays period, as a lot of people will not be at home. Also around Christmas, it may be difficult to contact people as they are probably too busy with other activities. Sending questionnaires by mail in the Christmas period is also not a good idea, as they may get lost in all other mails (Christmas/New Year cards).

The response rate also depends on the mode of data collection. Deploying interviewers for data collection usually leads to higher response rates. They can persuade people to participate in a poll. Therefore, response rates of face-to-face and telephone polls are often higher than those of mail and online polls.

In the United States, response rates of opinion polls have fallen dramatically over the years. Pew Research Center (2012) reports that the response rate of a typical RDD telephone poll dropped from 36% in 1997 to 9% in 2012. According to Vavrek and Rivers (2008), response rates have deteriorated over time so that most media polls (telephone polls using RDD) have response rates of around 20%. Response rates of RDD telephone polls also dropped in the United Kingdom to around 20%, and in inner cities, they are even down to 10%.

Online polls are not successful in achieving high response rates. Generally, response rates for this mode of data collection are at most 40%. For example, Bethlehem and Biffignandi (2012) describe an experiment with a survey (the Safety Monitor), where the online mode resulted in a response of 41.8%.

Polls are not mandatory, but some surveys are. Mandatory surveys have higher response rates. An example is the American Community Survey (ACS), which is conducted by the U.S. Census Bureau. Respondents are legally obliged to answer all the questions, as accurately as they can. Failure to do so may result in a fine. In the end, the response rate of this survey is over 95%. This is a very high response rate.

A factor having an impact on the response rate is the length of the questionnaire. The more questions the respondents have to answer, and the longer it will take to complete the questionnaire, the higher the risk people refuse to cooperate. And even if they start, there is a risk of breaking off in the middle of the questionnaire. So the general advice is to keep the questionnaire as short as possible.

In practice, it is usually not possible to compute the nonresponse bias. After all, the bias depends on a comparison of the answers of the respondents with those of the nonrespondents. Some indication of the possible risks of nonresponse can be obtained by computing the worst case: how large can the bias at most be?

Here is an example. Suppose, an election poll was carried out, and the response rate was only 40%. Of the respondents, 55% said they will vote at the election. If 40% responded, 60% did not respond. There are two extreme situations. The first one is that all nonrespondents will not vote. Then, the percentage of voters in the complete sample would be equal to

$$0.40 \times 55\% + 0.60 \times 0\% = 22\%$$

The second extreme situation is that all nonrespondents will vote. Then, the percentage of voters in the complete sample would be equal to

$$0.40 \times 55\% + 0.60 \times 100\% = 82\%$$

TABLE 7.1    Bandwidth of the Estimator Due to Nonresponse

| Percentage in Response | Response Rate | | | |
|---|---|---|---|---|
| | 20 | 40 | 60 | 80 |
| 10 | 2–82 | 4–64 | 6–46 | 8–28 |
| 20 | 4–84 | 8–68 | 12–52 | 16–36 |
| 30 | 6–86 | 12–72 | 18–58 | 24–44 |
| 40 | 8–88 | 16–76 | 24–64 | 32–52 |
| 50 | 10–90 | 20–80 | 30–70 | 40–60 |
| 60 | 12–92 | 24–84 | 36–76 | 48–68 |
| 70 | 14–94 | 28–88 | 42–82 | 56–76 |
| 80 | 16–96 | 32–92 | 48–88 | 64–84 |
| 90 | 18–98 | 36–96 | 54–94 | 72–92 |

So, the estimate for the percentage of voters (55%) could also have been any other percentage between 22% and 82%. The bandwidth is very large. Indeed, the effects of nonresponse can be substantial. In fact, one must conclude that the outcomes of this poll are not meaningful.

Table 7.1 contains the bandwidth of the complete sample percentages for a series of response rates. It is clear that the bandwidth decreases as the response rate increases. For response rates below 60%, the bandwidth can be very large. For example, if the response rate is 60% and 50% of the respondents have a certain property, the full sample percentage is somewhere between 30% and 70%. These two percentages denote extreme situation which probably will not occur in practice. Nevertheless, it shows that the response percentage of 50% should not be taken for granted.

Nonresponse can have various causes. It is a good idea to classify these causes. Different causes of nonresponse may have different effects on estimators and therefore require different treatment. The three main causes of nonresponse are *noncontact* (NC), *refusal*, and *not-able*.

The first step in getting response from a person in the sample is establishing contact. If contact is not possible, the result is a case of nonresponse due to NC. There can be various reasons why the contact attempt fails. For example, someone is not at home, because he was away for a short period of time (shopping), for a long period of time (a long holiday), or even permanently (moved to a different, unknown, and address). NC in a face-to-face poll may also be caused by gatekeepers, such as guards or doormen of secured residential complexes and by dangerous dogs around the house. A selected person may even have deceased between the moment of sample selection and the contact attempt.

NC occurs in telephone polls if someone does not answer the telephone, or the telephone is busy. NC in mail polls may be caused by people being away for a long time, by using a wrong address, or by people throwing away all unwanted mail immediately. Moreover, NC in online polls occurs when the e-mail with the invitation does not reach a selected person, for example, because it does not succeed in passing a spam filter.

Nonresponse due to NC can be reduced by making several contact attempts. Someone who is away today can be at home tomorrow. It is not uncommon for large survey organizations, such as national statistical institutes, to make up to six contact attempts before the case is closed as a NC.

As soon as contact is established with people, it must be checked whether they belong to the target population. Those who do are called *eligible*. If people are not eligible, they do not have to be interviewed. They are cases of overcoverage, which means they can be ignored. If people are eligible, they have to be persuaded to complete the questionnaire. If the researcher fails to get cooperation, it is a case of nonresponse due to *refusal*.

It helps to distinguish temporary refusal and permanent refusal. Temporary refusal often means that date and time are not suitable for answering the questions. The baby is crying, one is busy in the kitchen, or there is a football match on TV. Maybe it is possible to make an appointment for a later date and time. In case of a permanent refusal, it will not be possible to get the answers at any time. Permanent refusal may, for example, occur if people do not like the topic of the poll or if they consider the poll an intrusion of their privacy.

Even if it is possible to contact people and they want to participate, it still may be impossible to obtain their answers to the questions. Physical or mental conditions may prevent them doing so. This is called nonresponse due to *not-able*. People can be ill, drunk, deaf, blind, or have a mental handicap. These conditions can be temporary (which means another attempt can be made at another moment) or permanent.

Another condition causing people not to fill in the questionnaire is a language problem. They speak a different language, and therefore, they do not understand the questions. The latter problem can be solved by having multilingual interviewers or having copies of the questionnaire in different languages.

Table 7.2 gives an example of an overview of the response and nonresponse in a poll. A table like this one should always be included in the

TABLE 7.2   Fieldwork Results of the Survey on
Well-being of the Population (1998)

| Outcome | Frequency | Percentage |
|---|---|---|
| Response | 24,008 | 61.1 |
| Nonresponse | 15,294 | 38.9 |
| • No-contact | 2093 | 5.3 |
| • Refusal | 8918 | 22.7 |
| • Not-able | 1151 | 2.9 |
| • Other | 3132 | 8.0 |
| Total | 39,302 | 100.0 |

poll documentation. The table describes the results of the Survey on Well-being of the Population that was conducted by Statistics Netherlands in 1998. It was taken from Bethlehem et al. (2011).

The response rate of the survey is 61.1%. A look at the various causes of nonresponse shows that *Refusal* is by far the largest nonresponse category. The category *Other* consists of cases that were not taken up by the interviewers, because other cases took too much time. The workload of the interviewers simply turned out to be too large. Other cases took too much time.

To be able to compare the response rates of different surveys and polls, it is important to have a standardized procedure for computing them. It is not always easy what to include under response and what under non-response. AAPOR (2016) has published a document with definitions of response rates. One definition of the response rate is

$$RR1 = \frac{I}{(I+P)+(NC+R+NA+O)}$$

where:

$I$ is the cases of complete response (interviews)

$P$ is the cases of partial response (questionnaire is only partially completed)

$NC$ is the cases of noncontact.

$R$ is the cases of refusal.

$NA$ is the cases of not-able.

$O$ is the other cases of nonresponse

The response rate in Table 7.2 is obtained by taking $I = 24,008$, $P = 0$, $NC = 2093$, $R = 8918$, $NA = 1151$, and $O = 3132$.

Note that response rate definition assumes partially completed questionnaires to be cases of nonresponse, which means that the data on the forms are not used. One could also decide to see partially completed forms as cases of response. This leads to a different definition $RR2$ of the response with $(I + P)$ in the nominator instead of $I$.

Also note that noneligible cases are ignored in the definitions $RR1$ and $RR2$. Only eligible cases count. This makes it difficult to determine the number of NCs. As there is no contact with these people, the researcher does not know how many of them are eligible and how many are not. Based on earlier experience, it is sometimes possible to estimate the number of eligible NCs.

## 7.3 NONRESPONSE ANALYSIS

If there is nonresponse in a poll, one should be careful with the interpretation of the outcomes. Nonresponse may cause estimates to be biased. Therefore, the estimation techniques of Chapter 6 do not work anymore. The survey data have to be corrected for a possible bias due to nonresponse.

A first step after data collection should always be carrying out a nonresponse analysis. This analysis should provide inside in how selective the nonresponse is. If this is the case, a correction has to be carried out. This correction is called *adjustment weighting*. This section is about nonresponse analysis, and Section 7.4 about nonresponse correction.

To explore the effect of nonresponse, often a model is used in which each person is assigned a certain, unknown response probability. This response probability is denoted by the Greek letter $\rho$ (rho). Persons with a high response probability ($\rho$ close to 1) often participate in surveys and polls, whereas persons with a low response probability ($\rho$ close to 0) rarely participate. The response probabilities of all people in the target population are denoted by $\rho_1, \rho_2, \ldots, \rho_N$, where $N$ is the size of the population.

Suppose, a simple random sample is selected from this population. Due to nonresponse, not everyone in the sample participates. Therefore, not the $n$ planned observations are obtained, but less. The number of respondents is denoted by $m$ (where $m$ is smaller than $n$).

Suppose, the variable $Y$ is measured in a poll, and it can assume two values: 1 of a person has a specific property (e.g., votes for a certain party) or 0 if this person does not have it (does not vote for this party). Also assume, objective of the poll is estimating the percentage $P$ of people having the specific property.

In the case of full response, the analogy principle applies. This means that the sample percentage $p$ is an unbiased estimator of the population percentage $P$. In the case of nonresponse, however, the analogy principle does not apply. The response is not representative anymore. People with a high response probability are overrepresented, and those with a low response probability are underrepresented. Consequently, the response percentage $p$ is biased. There is a systematic difference between the estimate and the true value. The bias of this estimator is equal to:

$$\text{Bias}(p) = \frac{\text{Correlation}(Y,\rho) \times \text{Standard deviation}(\rho) \times \text{Standard deviation}(Y)}{P\,\text{mean}(\rho)}$$

The mathematical details of this expression can, for example, be found in Bethlehem et al. (2011, Chapter 2). *Correlation* $(Y,\rho)$ is the correlation between the values of the target variable $Y$ and the response probabilities $\rho$. The correlation takes a value between $-1$ and $+1$. A value close to $-1$ or $+1$ indicates a strong correlation between the target variable and the response probabilities. The closer the correlation is to 0, the weaker the relationship. A value of 0 means no relationship at all. So, the stronger the relationship between $Y$ and $\rho$, the larger the bias.

A typical example is an election poll in which people are asked whether they are going to vote or not. There is often a reasonably strong relationship between this variable and the response probabilities: people with a high response probability also tend to vote, and those with a low response probability tend to refrain from voting. Therefore, the estimate of the percentage of voters will be biased. As voters are overrepresented, the estimate will be too high.

Another example is described in Groves et al. (2006). They describe experiments in which the topic of a survey is emphasized. This leads to more respondents interested in the topic and to less respondents not interested. As a consequence, the correlation between the target variables and the response probabilities is stronger and therefore the bias larger.

*Standard deviation* $(\rho)$ is the standard deviation of the response probabilities. It measures the amount of variation of the response probabilities. The more the response probabilities vary (there are people with low response probabilities, and with high response probabilities), the larger the standard deviation will be. If all response probabilities are the same, the standard deviation is 0, and the bias vanishes.

Note that the experiments of Groves et al. (2006) also causes the response probabilities of interested people to be larger and those of noninterested people to be smaller. This increases the variation of the response probabilities. This will also increase the bias.

Pmean ($\rho$) is the population mean of the response probabilities. This quantity can be estimated by dividing the number of respondents $m$ by the initial sample size $n$. If the response probabilities are small, the bias will be large. A low response rate means a low value of Pmean ($\rho$), and thus a risk of a large bias.

In conclusion, it can be said that a strong correlation between the target variables and the response probabilities is bad, different response probabilities are bad, and low response probabilities are bad.

How can one detect whether nonresponse is selective? The available data with respect to the target variables of the poll will not be of much use. There are only data available for the respondents and not for the nonrespondents. So, it is not possible to find out whether respondents and nonrespondents behave differently with respect to these variables. The way out for this problem is to use auxiliary variables. *Auxiliary variables* are variables that are measured in the poll and for which also their distribution in the population is known. For example, if 60% of the respondents is male, and 40% is female, whereas in the target population, 49% is male and 51% is female, one can conclude that there is something wrong. There are too many males in the poll and too few females. To say it in other words, the response is not representative with respect to the variable gender. Males respond better than females. Apparently, there is a relationship between gender and response behavior. This leads to *selective response*.

It is important to look for auxiliary variables that have a relationship with response behavior. If such variables are found, the response of the poll is selective with respect to these variables. If these auxiliary variables also have a strong relationship with the target variables of the poll, the estimates for the target variables are biased.

If it turns out the response is selective, the estimation procedures described in Chapter 6 cannot be used anymore. First, the researcher has to correct the data for the lack of representativity. He needs the auxiliary variables for this correction, particularly the variables that show a relationship with response behavior.

Where to find auxiliary variables? Variables are needed that are not only measured in the poll, but for which also the population distribution is available. Here are a few sources of auxiliary variables:

- The sampling frame. Some sampling frames contain more variables than just contact information. A good example is a population register. The population register of The Netherlands contains gender, age (derived from date of birth), marital status, and country of birth.

- The national statistical institute. National statistical institutes usually have the population distribution of many variables. But be careful! These variables must have been defined in the same way, and they also must have been measured for the same population as the variables in the poll.

- Observations by interviewers. Interviewers can observe some properties of the persons in the sample, such as the socioeconomic status of the neighborhood, the type of the house, and the age of the house.

Figure 7.5 contains an example of a graph that gives insight in a possible relationship between the auxiliary variable *degree of urbanization* and response behavior. The data come from a survey of Statistics Netherlands, the 1998 survey of well-being of the population. It is clear that the response rate is very low (between 40% and 50%) in strongly urbanized areas. These are the big cities. The less urbanized an area, the higher the response rate is. Indeed, the response rate in rural areas is almost 70%. The low response rate in big cities is a problem in many countries.

Figure 7.6 shows a graph for another variable: the size of the household of a person. Here too, there is a clear pattern. The response rate increases

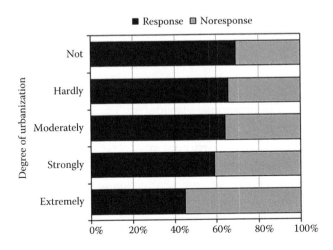

FIGURE 7.5   Response rate by degree of urbanization.

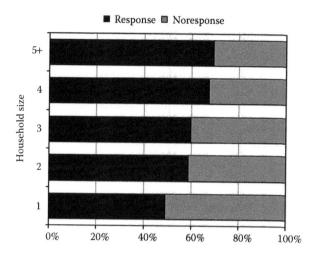

FIGURE 7.6   Response rate by household size.

as the size of the household increases. The main cause of the low response rate of one-person households is that these singles are very difficult to contact. Apparently, they are often away from home. Moreover, some small households (older singles and couples) are not able to participate. Finally, singles tend the refuse more than people in multiperson households. This all contributes to a lower response rate for one-person households.

This simple analysis shows that people in big cities and people in small households may be underrepresented in the poll. If something is investigated that is related to these variables, the researcher may expect estimates to be biased.

There are much more auxiliary variables that are related to response behavior in this survey. See Bethlehem et al. (2011) for a detailed treatment.

## 7.4 NONRESPONSE CORRECTION

If the analysis of nonresponse provides sufficient evidence for a potential lack of representativity, it is not scientifically sound to use the estimation procedures of Chapter 6 without any correction. Estimates would be biased. The most frequently used correction method is *adjustment weighting*. This comes down to assigning a *correction weight* to each respondent. In the computation of estimates, each respondent does no longer count for 1, but for the associated correction weight. Persons in underrepresented groups get a weight larger than 1, and persons in overrepresented groups get a weight smaller than 1.

To compute correction weights, auxiliary variables are required. Adjustment weighting is only effective if the auxiliary variables satisfy the following two conditions:

- They must have a strong relationship with the target variables of the poll. If there is no relationship, weighting will not reduce the bias of estimates for the target variables.

- They must have a strong relationship with the response behavior in the poll. If the auxiliary variables are unrelated with response behavior, adjustment weighting will not remove or reduce an existing response bias.

The idea of adjustment weighting is to make the response representative with respect to auxiliary variables. This is accomplished by computing correction weights in such a way that the weighted distribution of each auxiliary variable in the response is equal to the corresponding population distribution. The correction weights for underrepresented groups will be larger than 1, and those for overrepresented groups will be smaller than 1.

If it is possible to make the response representative with respect to a set of auxiliary variables and all these auxiliary variables have a strong relationship with the target variables, the weighted response will also become (approximately) representative with respect to the target variables. Therefore, estimates based on the weighted response will be better than estimates based on the unweighted response.

The elections in the fictitious town of Rhinewood are used to give a simple example of adjustment weighting. The number of eligible voters is 30,000. An election poll is carried out. The initial sample size is 1000, but due to nonresponse, only 500 respondents remain. Gender could be used as an auxiliary variable, because the gender of the respondents was recorded in the poll, and the population distribution of gender in the town is known. Table 7.3 shows how the weights are computed.

TABLE 7.3    Adjustment Weighting with the Variable Gender

| Response | | | Population | | | Correction Weight | |
|---|---|---|---|---|---|---|---|
| Gender | Frequency | Percentage | Gender | Frequency | Percentage | Gender | Weight |
| Male | 240 | 48.0 | Male | 15,330 | 51.1 | Male | 1.064583 |
| Female | 260 | 52.0 | Female | 14,670 | 48.9 | Female | 0.940385 |
| Total | 500 | 100.0 | Total | 30,000 | 100.0 | | |

The percentage of males in the response (48.0%) differs from the percentage of males in the population (51.1%). There are too few males in the response. The response can be made representative with respect to gender by assigning all males a weight equal to

$$\frac{\text{Percentage of males in the population}}{\text{Percentage of males in the response}} = \frac{51.1}{48.0} = 1.064583$$

In a similar way, all females get a correction weight equal to

$$\frac{\text{Percentage of females in the population}}{\text{Percentage of females in the response}} = \frac{48.9}{52.0} = 0.940385$$

Not surprisingly, the correction weight of males is larger than one (1.064583). Indeed, they are underrepresented in the response. After adjustment weighting, each male counts for 1.064583 male. Females are overrepresented and get a weight smaller than 1 (0.940385). Hence, each female will count for only 0.940385 female.

Suppose that the weighted response is used to estimate the percentage of males in town. All 240 males have a weight of 1.064583. So they weighted percentage is

$$100 \times \frac{240 \times 1.064583}{500} = \frac{255.5}{500} = 51.1$$

This is exactly the percentage of males in the target population. Likewise, the estimate for the percentage of females in the target population is exactly equal to the true population value. So the conclusion can be that the weighted response is representative with respect to gender.

Until now, adjustment weighting with one auxiliary variable was described. It is better to use more auxiliary variables, as this is more effective to reduce nonresponse bias. Adjustment weighting with several auxiliary variables is a bit more cumbersome, as these variables have to be crossclassified. It is shown how this works with an example for two auxiliary variables gender and age (in three categories young, middle-aged, and old). In the case of one auxiliary variable, there are as many groups (also called *strata*) as the variable has categories. So, for gender, there are two groups (males and females). In the case of two auxiliary variables gender and age, there is a group for each combination of gender and age. So, there

TABLE 7.4    Adjustment Weighting with Two Auxiliary Variables

| Response | | | Population | | | Correction Weight | | |
|---|---|---|---|---|---|---|---|---|
| Age | Male | Female | Age | Male | Female | Age | Male | Female |
| Young | 115 | 75 | Young | 6780 | 6270 | Young | 0.982609 | 1.393333 |
| Middle | 80 | 85 | Middle | 4560 | 4320 | Middle | 0.950000 | 0.847059 |
| Old | 65 | 80 | Old | 3990 | 4080 | Old | 1.023077 | 0.850000 |

are 2 × 3 = 6 groups (young females, middle-aged females, old females, young males, middle-aged males, and old males). If the distribution in the target population is known over these six groups, a correction weight can be computed for each group. Table 7.4 illustrates this with a numerical example.

The weights were computed in exactly the same way as in the example with one auxiliary variable. Weights are equal to population percentages divided by corresponding response percentages. For example, the percentage of old females in the population is 100 × 4080/3000 = 13.6%, whereas the percentage of old females in the response is 100 × 80/500 = 16.0%. So the correction weight for this group is equal to 13.6/16.0 = 0.850000.

As a result of weighting by gender and age, the response becomes representative with respect to both gender and age. Moreover, the response becomes representative with respect to gender within each age group, and it becomes representative with respect to age for each gender.

As more and more relevant auxiliary variables are used, weighting will be more effective in reducing the nonresponse bias. Keep in mind, however, that adjustment weighting only works if the groups obtained by crossing auxiliary variables, satisfy the two conditions already mentioned. They are rephrased here:

- The groups have to be homogeneous with respect to the target variables in the poll. This means that the persons in the groups must resemble each other with respect to the target variables. To say it in a different way, the values of a target variable must vary between groups and not within groups.

- The groups have to be homogeneous with respect to response behavior. This means that the persons in the groups must have more or less the same response probabilities. To say it in a different way, the values of a response probability must vary between groups and not within groups.

It is not always easy in practice to find proper auxiliary variables for adjustment weighting. Often, the researcher simply has to do with the variables he has available. As a consequence, nonresponse correction will be less effective. The bias may have reduced somewhat, but it will not vanish completely.

The application of adjustment weighting is illustrated by an example of a real poll. It was a radio-listening poll conducted in a town in The Netherlands in 2004. The main research question was how many people were listening to the local radio station. The target population consisted of all 19,950 inhabitants of the town with an age of 12 years and over.

A simple random sample of addresses was selected, and at each selected address, one person was drawn at random. So it was a two-stage sample. The initial sample consisted of 499 persons. In the end, 209 persons participated in the poll. So the response rate was $100 \times 209/499 = 41.9\%$. This is a low response rate. This means that there was a serious risk of biased estimates.

For the population of age 12 and older, the distribution over gender and age (in 3 categories) was available. Gender and age were recorded in the poll, so that it was possible to apply adjustment weighting by gender and age. Table 7.5 contains the data.

There are substantial differences between the response distribution and the population distribution. Females are overrepresented in all age groups. Middle-age males are clearly underrepresented. They are probably hard to contact, because they are working. Old males are heavily overrepresented.

Note that an address sample was drawn. This implies that people have unequal selection probabilities. The weights in Table 7.5 are correct for both the unequal selection probabilities as well as well as for unequal response probabilities. The effect of adjustment weighting on the estimates of the percentage of people listening to the local radio station is summarized in Table 7.6.

TABLE 7.5    Weighting the Radio-Listening Poll

| Response ($n = 209$) | | | Population ($n = 19,950$) | | | Weight | | |
|---|---|---|---|---|---|---|---|---|
| Age | Male (%) | Female (%) | Age | Male (%) | Female (%) | Age | Male | Female |
| Young | 9.5 | 18.3 | Young | 12.5 | 13.0 | Young | 1.136 | 0.613 |
| Middle | 17.7 | 28.6 | Middle | 26.3 | 27.7 | Middle | 1.284 | 0.835 |
| Old | 13.7 | 12.3 | Old | 8.9 | 11.6 | Old | 0.561 | 0.817 |

TABLE 7.6    Estimating the Percentage of Listeners

| Do You Ever Listen to the Local Radio Station? | |
| --- | --- |
| Estimate based on uncorrected response | 55.0% |
| Estimate after correction for unequal selection probabilities | 59.1% |
| Estimate after correction for nonresponse | 57.8% |

The uncorrected estimate is 55.0%. This is a wrong estimate as it is not corrected for unequal selection probabilities and nonresponse. The second estimate (59.1%) only corrects for unequal selection probabilities. This would be a correct estimate in the case of full response. The third estimate (57.8%) corrects for both unequal selection probabilities and nonresponse. This is the best estimate.

If there are many auxiliary variables, it may become difficult to compute correction weights. There could be groups (strata) without observations, in which it is simply not possible to compute weights (division by zero). Moreover, the population distributions of the crossing of variables may be missing. Other, more general, weighting techniques can be applied in these situations, such as *linear weighting* (*generalized regression estimation*) or *multiplicative weighting* (*raking ratio estimation*). See Bethlehem et al. (2011, Chapter 8) for more information.

Another weighting technique is gaining more and more popularity. It is called *propensity weighting*. Participation in the poll is modeled by means of a logistic regression model. This comes down to predicting the probability of participation in a poll from a set of auxiliary variables. The estimated participation probabilities are called *response propensities*. A next step could be to form groups of respondents with (approximately) the same response propensities. Finally, correction weight can be computed for each group. A drawback of propensity weighting is that the individual values of the auxiliary variables for the nonparticipating persons are required. Such information is often not available. Note that response propensities can also be used in other ways to reduce the bias (Bethlehem et al., 2011; Bethlehem, 2010).

## 7.5 SUMMARY

Even if a poll is well designed, with a good questionnaire, and with data collection based on a random sample, still then things can go wrong. Nonresponse is one of those things. It is the phenomenon that people in the selected sample do not provide the requested information. The three main causes of nonresponse are noncontact, refusal, and not-able.

Of course, nonresponse causes less data to becoming available. More importantly, nonresponse can be selective. It can cause specific groups to be under- or overrepresented in the survey response. Consequently, estimates will be biased. The magnitude of the bias depends on several factors.

A low response rate can lead to a large bias. Therefore, it is important to put efforts in keeping the response rate high. If the response rate is lower than, say, 50%, the researcher is often in for serious trouble.

The more the response probabilities of the people vary, the larger the bias will be. The ideal situation is the one in which everyone has the same response probability. Then there is no bias.

Realize that just increasing the sample size does not help to reduce the bias. Aiming at the *low hanging fruit*, that is, people in the response with a high response probability, does not solve the problem. The researcher should focus on people with low response probabilities. It is important to get those in the response. As this will decrease the variation in the response probabilities, it will also reduce the bias.

A third factor determining the magnitude of the nonresponse bias is the strength of the relationship between the target variables of the poll and the response probabilities. The stronger the correlation, the larger the bias. If there is no correlation, there is no bias. Unfortunately, correlations often occur.

The researcher must make every effort to keep the response rate at a high level. Nevertheless, there will almost always be nonresponse. To correct for a bias due to nonresponse, a correction must be carried out. This correction is called adjustment weighting. It comes down to assigning weights to all respondents in such a way that the response becomes representative with respect to a set of auxiliary variables, also called weight variables.

If the weight variables are correlated with the target variables and if the weight variables are correlated with response behavior (the response probabilities), adjustment weighting will remove or reduce the bias. Beware that weighting with uncorrelated weight variables will have no effect. So adjustment weighting is no guarantee for success.

Nonresponse is a serious problem. To assess the quality of a poll, one must always take a look at the nonresponse situation. The survey documentation should provide sufficient information to do this. At least, the response rate should be reported. If the response rate is below 70%, weighting adjustment is called for to avoid biased outcomes. If the response rate is 50% or lower, one should seriously worry about the outcomes. Adjustment weighting may not help to completely remove the bias. And if the response rate is below, say, 30%, one must wonder whether it is meaningful at all to take a serious look at the outcomes of the poll.

For each auxiliary variable, the poll documentation could contain a comparison of its response distribution with its population distribution. This gives some clues as to what the effects of nonresponse on the representativity of the poll could be.

If there is nonresponse, the researcher must have carried out some form of adjustment weighting. This should be reported in the poll documentation, and it must be clear which auxiliary variables were used for adjustment weighting.

# CHAPTER 8

# Online Polls

## 8.1 THE RISE OF ONLINE POLLS

When results of a poll are published, it is not always clear from the documentation how the data were collected. There are various modes of data collection possible, the most common ones being face-to-face, by telephone, by mail, and online. Each mode has its own advantages and drawbacks. To be able to assess the quality of a poll, it is therefore good to know more about the mode of data collection used.

One mode of data collection is becoming more and more popular, and this is online data collection. This is because online polls are an easy, fast, and cheap way to conduct a poll. Unfortunately, such polls are not always good polls. Online polls can have disadvantages that make their outcomes less reliable.

In view of the growing use of online polls, it is good to take a closer look at this type of poll. This is the objective of this chapter. The main message is that there are good and bad online polls, and that it is not always so easy to separate the wheat from the chaff.

Traditionally, polls used paper questionnaires to collect data. Data collection came in three modes: face-to-face, by telephone, and by mail. Developments in information technology in the 1970s and 1980s led to the rise of computer-assisted interviewing (CAI). The paper questionnaire was replaced by a computer program asking the questions. The computer took control of the interviewing process, and it also checked the answers to the questions on the spot. CAI could also be carried out in three different modes: computer-assisted telephone interviewing (CATI), computer-assisted personal interviewing (CAPI), and computer-assisted self-interviewing (CASI).

The rapid development of the internet in the last decades has led to a new mode of data collection: the online poll. This is sometimes also called *computer-assisted web interviewing.* Online polls became rapidly very popular. This is not surprising that as these polls seem to have some attractive properties which are as follows:

- Many people are connected to the internet. So an online poll is a simple means to get access to a large group of potential respondents.

- Online polls are cheap compared with other modes of data collection. No interviewers are required, and there are no mailing and printing costs.

- A poll can be launched very quickly. Just design a questionnaire and put it on the internet.

So, online polls seem to be an easy, cheap, and fast means of collecting large amounts of data. Everyone can do it! There are many tools on the internet, some of them even free, for setting up an online poll. Just go to such a website, enter some questions, and start the poll. Online polls, however, can also have serious methodological problems. There are sampling problems (how representative is the sample?) and measurement problems (does the poll measure what it was intended to measure?). If these problems are not seriously addressed, an online poll may result in low quality data for which the results cannot be generalized to the target population. This chapter discusses the following issues:

- *Undercoverage.* Can everyone in the target population participate in the poll? As not everyone in the target population may have access to the internet, portions of the population could be excluded from the poll. This may lead to biased estimates.

- *Sample selection.* How is the sample selected? Is it a simple random sample of persons having access to the internet? Or is it a case of self-selection of respondents, an approach which is known to produce bad poll outcomes.

- *Nonresponse.* Almost every poll suffers from nonresponse. An online poll is no exception. Nonresponse rates are particularly high for self-administered polls like mail and online polls. Unfortunately, low response rates can increase systematic error in the outcomes.

- *Measurement errors.* Interviewer-assisted polls like face-to-face polls and telephone polls are known to produce high-quality data. However, interviewer assistance is missing for online polls, and this may lead to more measurement errors. Questionnaire design is even more critical for online polls.

## 8.2 UNDERCOVERAGE IN ONLINE POLLS

Undercoverage occurs in a poll if the sampling frame only contains a part of the target population. If the sampling frame does not completely cover the target population, some people have a zero probability of being selected in the sample. If these people differ from those in the sampling frame, there is a serious risk of estimators being biased.

The obvious sampling frame for an online poll would be a list of e-mail addresses. Sometimes, such a sampling frame exists. For example, all employees of a large company may have a company e-mail address. Similarly, all students of a university usually have a university-supplied e-mail address. The situation is more complicated for polls having the general population as the target population. In the first place, not everyone in this population has access to the internet, and in the second place, there is no list of e-mail addresses of those with internet.

Figure 8.1 shows *internet coverage* in a number of European countries in 2015. Internet access of households varies considerably. Internet coverage is highest in the north of Europe. The top three are Norway (97%), Luxemburg (97%), and The Netherlands (96%). Almost all people in these countries have internet access. Internet coverage is low in the south-east of Europe. The countries with the lowest internet access are Bulgaria (59%), Romania (68%), and Greece (68%). Most recent data about internet access in the United States are from 2013. According to File and Ryan (2014), the level of internet access in the United States was 74% at the time.

Note that having access to internet is no guarantee that it is also used. There are people, for example, the elderly, who may have an internet connection but are not very active. One reason could be they simply lack the skills to do polls. This is a form of nonresponse.

There would be no undercoverage problem if there were no differences between those with and without internet access. In this case, a random sample from the population with internet access would still

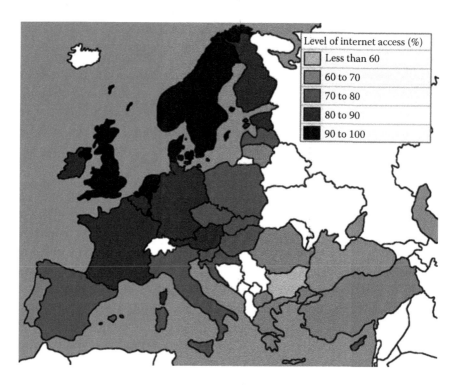

FIGURE 8.1 Percentage of households with internet access at home in 2015. (From Eurostat, Luxembourg, 2016.)

be representative of the whole population. Unfortunately, there are differences. Internet access is not equally spread across the population. Bethlehem and Biffignandi (2012) show that in 2005, the elderly, the low educated, and ethnic minority groups in The Netherlands were less well represented among those with internet access. Although internet coverage is very high in The Netherlands (96%), only 34% of the elderly (of age 75 years and older) use the internet (Statistics Netherlands, 2016).

Scherpenzeel and Bethlehem (2011) describe the LISS panel. This is a Dutch online panel recruited by means of probability sampling from a population register. Here too, the elderly and ethnic minority groups are underrepresented. Moreover, internet access among single households is much lower. They also conclude that general election voters are overrepresented.

Similar patterns can be found in other countries. For example, Couper (2000) describes coverage problems in the United States. It turns out that Americans with higher incomes are much more likely to have access to the internet. Black and Hispanic households have much less internet access than White households. People with a college degree have more internet access than those without it. Furthermore, internet coverage in urban areas is better than in rural areas.

Dillman and Bowker (2001) mention the same problems and conclude therefore that coverage problems of online polls of the general population cannot be ignored. Certain specific groups are substantially underrepresented. Specific groups in the target population will not be able to fill in the online questionnaire form.

Finally, Duffy et al. (2005) conclude that in the United States of America and the United Kingdom, online poll respondents tend to be politically active, are more likely to be early adopters of new technology, tend to travel more, and eat out more.

Suppose, the percentage of persons $P$ in a target population having a certain property is estimated. The number of people in this population is denoted by $N$. This target population can be divided into a population of size $N_I$ of people with internet access, and a population of size $N_{NI}$ of people without internet. So $N_I + N_{NI} = N$. If a simple random sample is selected from only the people with internet access, it can be shown that the sample percentage $p$ is a biased estimator of the population percentage $P$. According to Bethlehem and Biffignandi (2012), the undercoverage bias of this estimator is equal to

$$\text{Bias}(p) = \frac{N_{NI}}{N} \times \left( P_I - P_{NI} \right)$$

where:

$P_I$ is the percentage in the population of all people with internet

$P_{NI}$ is the percentage in the population of all those without internet

So the more these two percentages differ, the larger the bias will be. The bias vanishes if the two percentages are equal.

The bias also depends on the relative size $N_{NI}/N$ of the population without internet. The larger this relative size, the larger the bias. So be careful with an online survey that is conducted in a country with low internet coverage, such as the countries in the south-east of Europe. In the northern countries of Europe, internet coverage is so high that the bias is hardly a problem.

It is important to realize that the undercoverage bias of an online poll does not depend on the sample size. It is not uncommon that online polls have many respondents. But this does not guarantee the bias to be small. Moreover, increasing the sample size will have no effect on the bias. So, the problem of undercoverage in online polls cannot be solved by collecting more data.

The fundamental problem of online polls is that persons without internet are excluded from the poll. There are a number of ways in which a researcher could reduce the problem. The first approach is that of the LISS panel. This is a Dutch online panel consisting of approximately 5000 households. The panel is based on a true probability sample of households drawn from the population register of The Netherlands. Recruitment took place by CATI (if there was a listed telephone number available) or CAPI. People in households without internet access, or people who were worried about an internet poll being too complicated for them, were offered a simple to operate computer with internet access. The computer could be installed in their homes for free for the duration of the panel. So the size of the noninternet population was reduced by giving internet access to those without it.

An even more elegant solution is that of the ELIPSS panel in France. This is also an online panel. All members of the panel were provided with a touch-screen tablet (a Samsung Galaxy Tab 7) and a 3G internet-access subscription. This guarantees that all the respondents used the same device. Of course, this is not a cheap solution.

Another approach would be to turn the single-mode online poll into a *mixed-mode poll*. One simple way to do this is to send an invitation letter to the persons in the sample and give them a choice to either complete the questionnaire on the internet or on paper. So, there are two modes here: online and mail.

A researcher could also implement a sequential mixed-mode approach. First, people in the sample are offered the cheapest approach, and that is, online interviewing. They are invited to fill in the questionnaire on the internet. Nonrespondents are reapproached by telephone, if a listed telephone number is available. If not, these nonrespondents could be reapproached face-to-face.

It will be clear that these approaches increase the costs of a poll, and it also may increase the duration of the fieldwork. However, this may be the price a researcher has to pay for reliable statistics. More about the opportunities and challenges of mixed-mode polls can, for example, be found in Bethlehem and Biffignandi (2012).

## 8.3 SAMPLE SELECTION FOR AN ONLINE POLL

Chapter 5 already stressed the importance of selecting a sample for a poll by means of probability sampling. This makes it possible to compute valid (unbiased) estimates of population characteristics. Sample selection must be such that every person in the target population has a nonzero probability of selection, and all these probabilities must be known. Furthermore, probability sampling under these conditions makes it possible to compute the precision of estimates.

The principles of probability sampling also apply to online polls. For example, if a poll is conducted among students of a university, a list of e-mail addresses can be used as a sampling frame for a simple random sample. Unfortunately, many online polls, particularly those conducted by market research organizations, are not based on probability sampling. The questionnaire of the poll is simply put on the web. Respondents are those people who happen to have internet, visit the website, and spontaneously decide to participate in the poll. As a result, the researcher is not in control of the selection process. Selection probabilities are unknown. Therefore, no unbiased estimates can be computed, nor can the precision of estimates be determined. These polls are called here *self-selection polls*. Sometimes, they are also called *opt-in polls*.

Self-selection polls have a high risk of not being representative. Following are several phenomena that contribute to this lack of representativity:

- Often people from outside the target population can also participate in the poll. This makes it impossible to generalize the poll results to the intended target population.

- It is sometimes possible for respondents to complete the questionnaire more than once. Even if there is a check on the IP address of the computer of a respondent (the questionnaire can only be completed once on a specific computer), it is still possible for the same person to do the poll on another device (computer, laptop, tablet, or smartphone).

- Specific groups may attempt to manipulate the outcomes of the poll, by calling upon the members of the group to participate in the poll, and to complete the questionnaire in a way that is consistent with the objectives of the group.

Chapter 5 contains an example of online self-selection poll. The poll was carried out during the campaign for the 2014 local election in Amsterdam. Objective of the poll was to find out which party leader won the local debate. Two campaign teams discovered that the questionnaire could be completed more than once. So they stayed up all night and did the poll many times. Lacking representativity, the poll was cancelled the next morning.

Chapter 5 contains another example of a poll that was manipulated. An online self-selection poll was conducted to determine the winner of the 2005 Book of the Year Award. Voters could select one of the nominated books or use an open question to mention a different book of their choice. There was a campaign by Bible societies, a Christian broadcaster, and a Christian newspaper to vote for the non-nominated new Bible translation. The campaign was successful. An overwhelming 72% of the 92,000 respondents voted for this book. Of course, this poll was far from representative.

There was a case of attempted poll manipulation during the campaign for the general election in The Netherlands in 2012. A group of people tried to influence a major opinion poll. The group consisted of 2500 people. They intended to subscribe to this opinion panel. Their plan was to behave themselves first as Christian Democrats (CDA). Later on, they would change their opinion and vote for the elderly party (50PLUS). They hoped that this would trigger also other CDA voters to change to 50PLUS. Unfortunately for them, and fortunately for the researcher, their attempt was discovered when suddenly so many people at the same time subscribed to the panel. See Bronzwaer (2012) for more details.

Almost all political polls in The Netherlands are online polls that use self-selection for recruiting respondents. Therefore, the outcomes of these polls may be biased. This became clear during the general elections in 2012. The elections were preceded by a short, but intense, campaign in which polling organizations were very active. The four main ones were Maurice de Hond (*Peil.nl*), Ipsos (*Politieke Barometer*), TNS NIPO, and GfK Intomart (*De Stemming*). They conducted many polls. Some of them even did a poll every day in the final phase of the campaign. Table 8.1 compares the election results (seats in parliament) with the outcomes of the polls one day before the election. For all four polling organizations, there were significant differences between predictions and the real result. In other words, these differences were larger than the margin of error.

The largest difference was found for the prediction of the *Socialist Party* (*SP*) by the *De Stemming*. The prediction was 22 seats in parliament, but

TABLE 8.1  Predictions (Seats in Parliament) for the General Election in The Netherlands in 2012

| Party | Election Result | Peil.nl | Politieke Barometer | TNS NIPO | De Stemming |
|---|---|---|---|---|---|
| VVD (Liberals) | 41 | 36 | 37 | 35 | 35 |
| PvdA (Social-democrats) | 38 | 36 | 36 | 34 | 34 |
| PVV (Populists) | 15 | 18 | 17 | 17 | 17 |
| CDA (Christian-democrats) | 13 | 12 | 13 | 12 | 12 |
| SP (Socialists) | 15 | 20 | 21 | 21 | 22 |
| D66 (Liberal-democrats) | 12 | 11 | 10 | 13 | 11 |
| GroenLinks (Green) | 4 | 4 | 4 | 4 | 4 |
| ChristenUnie (Christian) | 5 | 5 | 5 | 6 | 7 |
| SGP (Christian) | 3 | 3 | 2 | 2 | 3 |
| PvdD (Animals) | 2 | 3 | 3 | 2 | 2 |
| 50PLUS (Elderly) | 2 | 2 | 2 | 4 | 3 |
| Total difference | | 18 | 18 | 24 | 24 |
| Mean difference | | 1.6 | 1.6 | 2.2 | 2.2 |

this party got only 15 seats in the real election. Four times a difference of six seats can be observed, and two times there is a difference of five seats.

These differences can at least partly be attributed to using self-selection for the polls. Another possible explanation, suggested by some polling organizations, was that people could have changed their opinion between the final poll and the election. This is sometimes called a *late swing*.

So, self-selection polls produce biased estimates. One can obtain insight in the magnitude of the bias by assuming that each person in the population has an unknown *participation probability*. If all participation probabilities are the same, there is no bias. Problems occur when these probabilities differ. The magnitude of the bias depends on the following three factors:

- The correlation between the participation probabilities and the target variables of the poll. The stronger the correlation, the larger the bias.

- The variation of the values of the participation probabilities. The more the values of these probabilities vary, the larger the bias will be.

- The mean of the participation probabilities. The smaller the mean, the larger the bias.

It will be clear that, due to the problems of self-selection, one cannot draw scientifically sound conclusions from an online self-selection poll. It is impossible to use such polls for drawing valid and reliable conclusions. Therefore, a poll should be based on a probability sample from a proper sampling frame.

Indeed, the American Association for Public Opinion Research warns for the risks of self-selection (Baker et al., 2010).

> Only when a web-based survey adheres to established principles of scientific data collection can it be characterized as representing the population from which the sample was drawn. But if it uses volunteer respondents, allows respondents to participate in the survey more than once, or excludes portions of the population from participating, it must be characterized as unscientific and is unrepresentative of any population.

Selecting a probability sample provides safeguards against various manipulations. It guarantees that sampled persons are always in the target population, and they can participate only once in the poll. The researcher is in control of the selection process.

In considering use of the internet for a poll, the ideal sampling frame would be to have a list of e-mail addresses of every person in the target population. Unfortunately, such sampling frames do not exist for general population polls. A way out is to do recruitment in a different mode. One obvious way of doing this is to send sample persons a letter with an invitation to complete the questionnaire on the internet. To that end, the letter must contain a link to the poll website, and also the unique identification code.

Recruitment by mail is more cumbersome than by e-mail. In case of e-mail recruitment, the questionnaire can simply be started by clicking on the link in the e-mail. In the case of mail recruitment, more actions are required. If one has a desktop computer, it could mean going to the computer (e.g., upstairs in de study room), starting the computer, connecting to the internet, and typing the proper internet address (with the risk of making typing errors). If one only has a smartphone, it means finding it, switching it on, connecting to the internet, and entering the internet address. So this is more work than just clicking on a link in an e-mail.

## 8.4 NONRESPONSE IN ONLINE POLLS

Like any other type of poll, online polls too suffer from nonresponse. Nonresponse occurs when persons in the selected sample, who are also eligible for the poll, do not provide the requested information. The problem of nonresponse is that the availability of data is determined both by the (known) sample-selection mechanism and the (unknown) response mechanism. Therefore, selection probabilities are unknown. Consequently, it is impossible to compute unbiased estimates. Moreover, use of naive estimators will lead to biased estimates.

Nonresponse can have several causes. It is important to distinguish these causes as different causes can have different effects on estimators, and therefore they may require different treatment. Three causes of nonresponse were already described in detail in Chapter 7: noncontact, refusal, and not-able.

*Noncontact* occurs when it is impossible to get into contact with a selected person. Various forms of noncontact are possible in an online poll. It depends on the way in which people were recruited for the poll. If the sampling frame is a list of e-mail addresses, there is noncontact when the e-mail with the invitation to participate in the poll does not reach the selected person. The e-mail address may be wrong, or the e-mail may be blocked by a spam filter. If the sampling frame is a list of postal addresses and letters with an internet address are sent to selected persons, noncontact may be caused by not receiving the letter. If recruitment for an online poll takes place by means of a face-to-face or telephone poll, noncontact can be due to respondents being not at home or not answering the telephone.

Nonresponse due to *refusal* can occur after contact has been established with a selected person. Refusal to cooperate can have many reasons: people may not be interested in the topic of the poll, they may consider it an intrusion of their privacy, they may have no time, and so on.

If sample persons for an online poll are contacted by an e-mail or a letter, they may postpone and forget to complete the questionnaire form. This can be seen as a weak form of refusal. Sending a reminder may help one to reduce this type of refusal.

Nonresponse due to *not-able* may occur if respondents are willing to respond but are not able to do so. Reasons for this type of nonresponse can be, for example, illness, hearing problems, or language problems. If a letter with an internet address of an online questionnaire is sent to selected persons, and they want to participate in the online poll but do not have access to the internet, this can also be seen as a form of nonresponse due to not-able.

Note that lack of internet access should sometimes be qualified as undercoverage instead of nonresponse. If the target population of a poll is wider than just those with internet, and the sample is selected using the internet, people without internet have a zero-selection probability. They will never be selected in the poll. This is undercoverage. Nonresponse due to not-able occurs if people have been selected in the poll but are not able to complete the questionnaire form (on the internet).

Nonresponse can cause the outcomes of an online poll to be biased. The magnitude of the bias is determined by several factors. To describe these factors, it is assumed that each person in the target population has a certain unknown probability of response when selected in the sample. The magnitude of the nonresponse bias depends on the following:

- The correlation between the response probabilities and the target variables of the poll. The stronger the correlation, the larger the bias.

- The variation of the values of the response probabilities. The more these probabilities vary, the larger the bias will be.

- The mean of the response probabilities. The smaller the mean (the lower the response rate), the larger the bias.

The bias vanishes if all response probabilities are equal. In this case, the response can be seen as a simple random sample. The bias also vanishes if the response probabilities do not depend on the value of the target variable.

Note that a poll with a low response rate will not be bad if all response probabilities are approximately equal. In a similar fashion, a poll with a high response rate can have serious problems if the response probabilities vary a lot.

The mode of data collection has an impact on the response rate of a poll. Typically, interviewer-assisted polls have higher response rates than self-administered polls. As an online poll is a self-administered poll, a researcher can expect the response rate to be lower than that of, for example, a face-to-face or a telephone poll. Indeed, the literature seems to show that this is the case. Cook et al. (2000) analyzed the response of 68 online polls. The average response rate was around 40%. Kaplowitz et al. (2004) compared response rates of online polls and mail polls. They concluded that these rates were comparable. They saw response rates vary between 20% and 30%, and Lozar Manfreda et al. (2008) compared the response of online polls with other types of polls. They found that, on average, the response rate of the online surveys was 11% lower than that of other surveys.

## 8.5 ADJUSTMENT WEIGHTING

Undercoverage, self-selection, and nonresponse can all affect the representativity of the response of an online poll. Consequently, estimators may be biased. To reduce these problems, some kind of correction procedure should be carried out. This is usually an adjustment-weighting technique.

The fundamental idea of adjustment weighting is to restore the representativity of the poll response. This can be accomplished by assigning weights to responding persons. People in underrepresented groups should have weights larger than 1, and those in overrepresented groups should have weights smaller than 1. Weights can only be computed if *auxiliary variables* are available. These variables must have been measured in the poll. Moreover, their population distribution must be available. By comparing the population distribution of an auxiliary variable with its response distribution, it can be assessed whether the response is representative with respect to this variable. If these distributions differ considerably, the response is selective.

Weighting adjustment is only effective if following two conditions are satisfied:

1. The auxiliary variables must have a strong correlation with the selection mechanism of the poll.

2. The auxiliary variables must be correlated with the target variables of the poll. If these conditions are not fulfilled, adjustment weighting is not effective. The estimators will still be biased.

The availability of a sufficient number of effective auxiliary variables is often a problem. Usually, there are not many variables that have a known population distribution and that satisfy the above two conditions. If proper auxiliary variables are not available, one could consider conducting a *reference survey*. The objective of such a survey is obtaining a good (unbiased and precise) estimate of the population distribution of relevant auxiliary variables. The (estimated) distributions can be used to compute correction weights.

Such a reference survey should be based on a real (possibly small) probability sample, in which data are collected with a mode different from the web, for example, face-to-face or by telephone. If there is full response, or if the nonresponse is unrelated to the auxiliary variables in the survey, the reference survey will produce unbiased estimates of the population distribution of these variables.

An interesting aspect of the reference survey approach is that any variable can be used for adjustment weighting as long as it is measured in both the poll and the reference survey. For example, some market research organizations use so-called *psychographic variables* to divide the population in *mentality groups*. People in the same group are assumed to have more or less the same level of motivation and interest to participate in polls. If this is the case, such variables can be effectively used for adjustment weighting.

The reference survey approach also has some disadvantages. In the first place, it is expensive to conduct an extra survey. However, this survey need not be very large as it is just used for estimating the population distribution of some auxiliary variables, and the information can be used for weighting more than one online poll. In the second place, Bethlehem (2010) shows that the variance of the weighted estimator is, for a substantial part, determined by the size of the (small) reference survey. So, a large number of observations in an online poll do not guarantee precise estimates. The reference survey approach reduces the bias of estimates at the cost of a much larger margin of error.

An example shows that adjustment weighting is often necessary in daily practice. As soon as a poll suffers from undercoverage, self-selection, or nonresponse, poll results may be biased. Therefore, one should always check whether adjustment weighting has been carried out.

There are three nationwide public TV channels in The Netherlands. One of these channels (NPO 1) has a current affairs program called *EenVandaag*. This program maintains an online panel. It is used to measure public opinion with respect to topics that are discussed in the program. The *EenVandaag Opinie Panel* started in 2004. In 2008, it contained approximately 45,000 members.

The panel is a self-selection panel. Participants are recruited among the viewers of the program. For these reasons, the panel lacks representativity. Therefore, it was explored how unbalanced the composition of the panel is, and whether estimates can be improved by applying weighting adjustment.

One of the waves of the opinion panel was conducted in April 2008 to determine preference for political parties. The questionnaire was

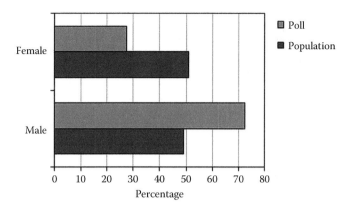

FIGURE 8.2   The distribution of gender in the poll and in the population.

completed by 19,392 members of the panel. The representativity of the response was affected by two phenomena. In the first place, the panel was constructed by means of self-selection. In the other place, not all members of the panel responded to the request to complete the questionnaire. The response rate was $100 \times 19{,}392/45{,}632 = 42.5\%$, which is not very high.

If persons apply for membership of the panel, they have to fill in a basic questionnaire with a number of demographic questions. This is often called a *profile poll*. The demographic variables measured can be used as auxiliary variables. The following variables were available: gender, age, marital status, province, ethnic background, and vote at the 2006 general election. Here, as an example, two variables are shown: gender and vote in 2006. Figure 8.2 compares the response distribution of gender with its population distribution. There are dramatic differences. More than 70% of the respondents in the poll is male, and only a little less than 30% is female. Both percentages should be close to 50%. So, males are overrepresented and females are underrepresented. Hence, estimates for any variable related to gender will be biased.

Figure 8.3 compares the response distribution of voting in the 2006 general election with its population distribution. Although it may not be clear where all the political parties stand for, the bar chart shows that there are substantial differences between both distributions. For example, the poll response contains too few people voting for the CDA. Voters for the SP are substantially overrepresented. There is also a dramatic difference for the nonvoters (the bars at the bottom of the graph). Five percent of the poll respondents indicated they did not vote, whereas the true percentage in

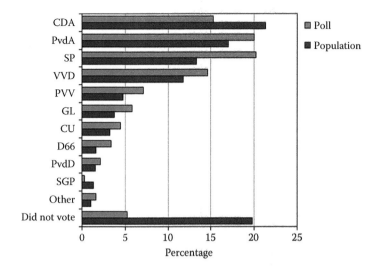

FIGURE 8.3    The distribution of voting in 2006 in the poll and in the population.

the population was around 20%. This phenomenon can often be observed in political polls. Voters are overrepresented, and nonvoters are underrepresented. There is a clear correlation between voting and participating in polls. People, who vote, also agree to participate in polls; and people, who do not vote, also refuse to participate in polls.

It is clear that both auxiliary variables (gender and voting) are correlated with response behavior. So they could qualify as weighting variables. Both variables were tried for one target variable which measures whether people intend to vote for the SP at the next general election. The results are summarized in Table 8.2. If no weighting is carried out, the estimate of the percentage of SP-voters is 12.9%. Weighting with gender has almost no effect. The shift in the estimate is very small (from 12.9 to 13.1). Apparently, there is no correlation between the auxiliary variable gender and the target variable SP. Weighting with voting in 2006 has a substantial effect. The estimated percentage of SP-voters drops from 12.9% to 9.8%, which is over three percentage points. One can conclude that weighting is

TABLE 8.2    The Effect of Weighting on the Estimate of the Percentage of Voters for the Socialist Party (SP)

| Weighting | Votes for SP (%) |
|---|---|
| No weighting | 12.9 |
| Gender | 13.1 |
| Voting in 2006 | 9.8 |

effective here. The effect is not surprising. Figure 8.3 shows that SP-voters in the 2006 election were over-represented. If this is corrected, less people remain that sympathize with the SP.

The conclusion is that some form of weighting adjustment must always be considered to reduce the bias of estimators. However, success is not guaranteed. The ingredients for effective bias reduction may not always be available. Be careful. Be suspicious if no adjustment weighting has been carried out in a poll.

## 8.6 MEASUREMENT ERRORS

Traditionally, many polls used face-to-face or telephone interviewing. These are not the cheapest modes of data collection, but they are used, because response rates are high, and data quality tends to be good. What would change in this respect if a face-to-face or telephone poll were to be replaced by an online poll? It would certainly have an impact on measurement errors. This section focuses on measurement errors in online polls and the effects they have on data quality. *Measurement error* is here defined as the difference between the recorded answer to a question and the correct answer.

Answering questions is not an easy task. Schwarz et al. (2008) describe the following steps the respondents have to go through: (1) understanding the question, (2) retrieving the required information from memory, (3) translating the information in the proper answer format, and (4) deciding whether to give the answer or not. One should realize a lot can go wrong in this process. This may particularly be a problem for online polls, in which there are no interviewers. So they cannot motivate respondents, answer questions for clarification, provide additional information, and remove causes for misunderstanding. Respondents decide these on their own.

When designing a questionnaire for an online poll, it should be realized that respondents are usually not interested in the topic of the survey. Therefore, participation is not important for them. Krug (2006) describes how people read websites. Many of his points also apply to questionnaires of online polls:

- Respondents do not read the text on the screen. They just scan it looking for words or phrases that catch the eye.

- Respondents do not select the optimal answer to the question, but the first reasonable answer. This is called *satisficing*.

- Respondents are aware of the fact that there is no penalty for giving wrong answers.

- Respondents do not read introductory texts explaining how the questionnaire works. They just muddle through and try to reach the end.

There are several ways in which online polls differ from face-to-face or telephone polls. One important aspect was already mentioned: *satisficing*. It means that respondents do not do all they can to provide a correct answer. Instead, they give a satisfactory answer with minimal effort. Satisficing is not only caused by bad behavior of respondents but also by badly designed questionnaires.

Online polls suffer more from satisficing than interviewer-assisted polls. Satisficing manifests itself in a number of ways. Some forms of satisficing in online polls are described in Sections 8.6.1 through 8.6.6.

## 8.6.1 Response Order Effects

Particularly, if the list of possible answers to a closed question is long, respondents tend to choose an answer early in the list. This is called the *primacy effect*. An example is the question in Figure 8.4. A respondent

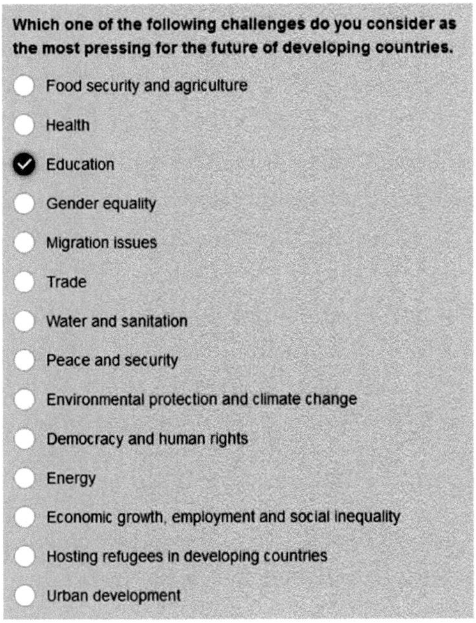

FIGURE 8.4  A closed question with many answer options.

may quickly loose interest and therefore pick an answer somewhere in the first part of the list.

Note that face-to-face and telephone polls may suffer from a *recency effect*. For this mode of data collection, interviewers read out the possible answers. When they reach the end of the list, respondents already have forgotten the first answers in the list, and therefore they will pick one of the last answers mentioned.

*Radio buttons* were used in Figure 8.4 to present the list of possible answers. This is a good way to format a closed question. Respondents have to select an answer, and they can only select one answer. As soon as they click on an answer, an already selected answer is deselected.

Particularly, for a long list of answer options, a closed question is sometimes formatted in a different way. Instead of radio buttons, a *drop-down list* is used. An example is shown in Figures 8.5 and 8.6. Figure 8.5 shows the question in its initial state. The list has not been unfolded yet, and no answer has been selected.

To answer this question, respondents first have to open the list by clicking on it. Next, they must select the proper answer from the list. Figure 8.6

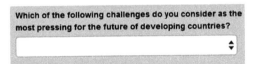

FIGURE 8.5   A closed question with a drop-down list (initial state).

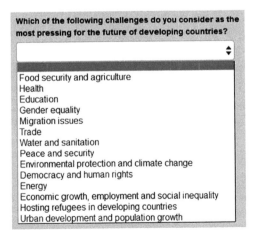

FIGURE 8.6   A closed question with a drop-down list (after opening the list).

FIGURE 8.7    A closed question with a long drop-down list.

shows the status of the list after it has been opened. So, answering a closed question with a drop-down list requires more actions than answering a closed question with radio buttons. Moreover, when respondents are confronted with a question similar to the one in Figure 8.5, it may be unclear to them what to do. What is expected from them? They may be confused.

The situation is even more confusing if the list is so long that it does not fit in the drop-down list window. An example is given in Figure 8.7. Initially, only the first part of the list is visible. Respondents may get the impression that these are all the possible answer options, and therefore they pick an answer from this visible part of the list. Hopefully, respondents notice the scrollbar that they can use to scroll through the list and make other parts of the list visible. If they fail to notice or use the scrollbar, they may easily pick an (incorrect) answer.

Research has shown that a drop-down list is not the best way to format a closed question, as such a question may suffer even more from primacy effects than a closed question with radio buttons. However, sometimes a drop-down list is hard to avoid if the list of possible answers is very long.

## 8.6.2 Endorsing the Status Quo

Respondents in online polls tend to endorse the status quo. If they are asked to give their opinion about changes, they simply (without thinking) select the answer to keep everything the same. Endorsing the status quo is typically a problem in self-administered polls.

An example of a question that may be affected is asking whether government should change its policy with respect to some issue (*Should gun control laws in the United States become more strict or less strict*).

According to Krosnick (1991), a small portion of respondents insists on the answer *no change* even if this option is not available. These are the real no-change respondents, and if this option is offered, the number of no-change respondents increases by 10%–40%. Many respondents see this answer is an easy way out of the question.

## 8.6.3 Selecting the Middle Option

Should a middle, neutral response option be offered in an opinion question? On the one hand, some researchers think it should be included, because there are people with a neutral opinion. These people must be able to say so. They may not be forced to give a different answer that does not correspond to their opinion. Figure 8.8 gives an example of a question with a middle option.

On the other hand, if a middle option is offered, it will be easy for respondents to select this option, without having to think about their real opinion. Experiments show that if the middle option is offered, many respondents will select it.

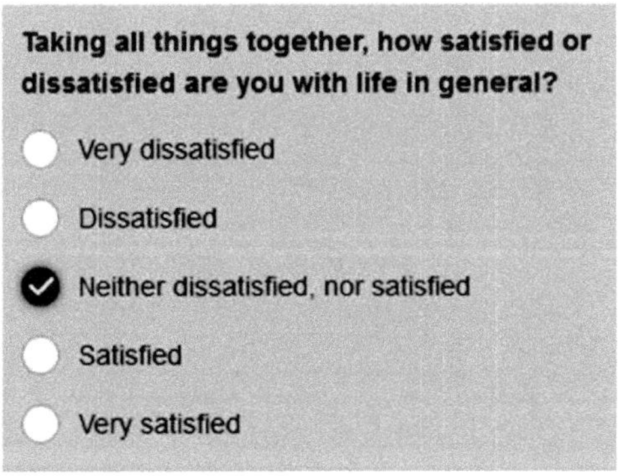

**Taking all things together, how satisfied or dissatisfied are you with life in general?**

○ Very dissatisfied

○ Dissatisfied

✓ Neither dissatisfied, nor satisfied

○ Satisfied

○ Very satisfied

FIGURE 8.8 A closed question with a middle answer option.

Unfortunately, there is no easy answer to the question about including a middle option. Nevertheless, some experts tend to lean to the opinion that it is better to have a middle option. For example, Bradburn et al. (2004) note that this often does not affect the ratio of positive to negative answers. Moreover, it produces more information about the intensity of opinions.

### 8.6.4 Straight-Lining

If there is a series of questions with the same set of possible answers, they can be combined into a *grid question* (sometimes also called a *matrix question*). A grid question has the format of a table in which each row represents a single question and each column corresponds to one possible answer option. Figure 8.9 shows a simple example of a grid question. In practice, grid questions are often much larger than this one.

At first sight, grid questions seem to have some advantages. They take less space in the questionnaire than a set of single questions. Moreover, it provides respondents with more oversight, making it easier to answer the questions.

Some experts have concerns about grid questions in online polls. Answering a grid question is more difficult (from a cognitive point of view) than answering a set of separate single questions. This may lead to *straight-lining*. Respondents take the easy way out by checking the same answer for all questions in the grid. They simply check all radio buttons in the same column.

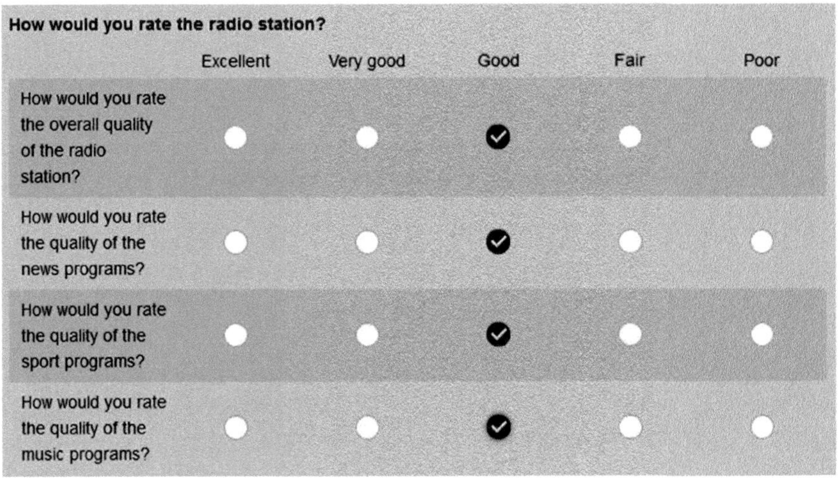

FIGURE 8.9  A grid question with a middle answer option.

Often this is the column corresponding to the middle (neutral) response option. Figure 8.9 contains an example of straight-lining.

It might be wise to limit the use of grid questions in online polls as much as possible, and if they are used, the grid should not be too wide or too long. Preferably, the whole grid should fit on a single screen. This is not so easy to do this as different respondents may have set different screen resolutions on their computer screens or use different devices (laptop, tablet, and smartphone). If they have to scroll, either horizontally other vertically, they may easily get confused, leading to wrong or missed answers.

### 8.6.5 Don't Know

There is a dilemma for handling *don't know* in online polls. On the one hand, this option should be explicitly offered as persons may really know they do not know the answer to a question. Some people have an opinion, and some have not. An example is shown in Figure 8.10. On the other hand, the option *don't know* may be unlikely for factual questions, such as questions about gender or age.

If the option *don't know* is available, the researcher runs the risk that many persons will select it as an easy way out. This can be seen as a form of satisficing.

In the case of a CAPI or CATI poll, it is also possible to offer *don't know* implicitly. The interviewer only reads out the substantive options, but if the respondent insists he does not know the answer, the interviewer can record the answer as *don't know*. This works well, but it is difficult to implement in an online poll.

Sometimes, respondents may avoid *don't know*, because they do not want to admit they do not know the answer. In this case, it may help us to include a *filter question*. This question asks the respondents first whether they have

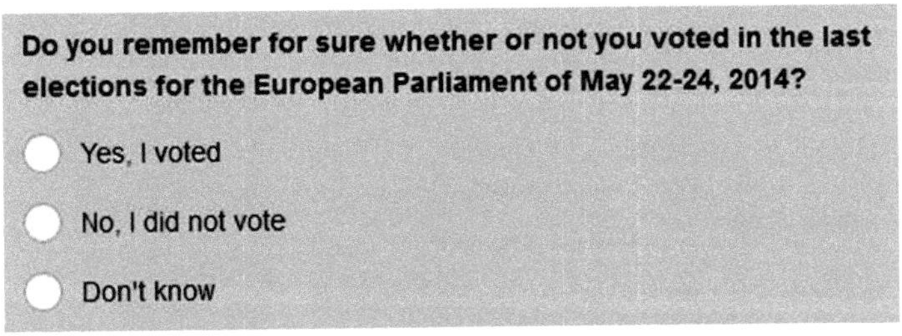

**Do you remember for sure whether or not you voted in the last elections for the European Parliament of May 22-24, 2014?**

- ◯ Yes, I voted
- ◯ No, I did not vote
- ◯ Don't know

FIGURE 8.10  A closed question with the option *don't know*.

an opinion on a certain issue. Only if they say they have, they are asked to specify their opinion in the subsequent question. See Krosnick (1991) and Schuman and Presser (1981) for a discussion about filter questions.

### 8.6.6 Arbitrary Answer

Respondents not wanting to think about the proper answer may decide to pick just an arbitrary answer. Krosnick (1991) calls this behavior of respondents *mental coin flipping*. It typically occurs for check-all-that-apply questions. It is typically a problem in online polls and mail polls.

Figure 8.11 contains an example. Checking all answers that apply may be a lot of work. Instead of checking all relevant answers, respondents can decide to check only a few arbitrary answers and stop when they think they have given enough answers. Moreover, satisficing respondents tend to read only the answers in the first part of the list, and not the complete list. This behavior may also lead to a primacy effects.

So one should be careful with check-all-that-apply questions with long lists of answer options. Note that a different format can be used for such a question. See Chapter 3 about questionnaire design.

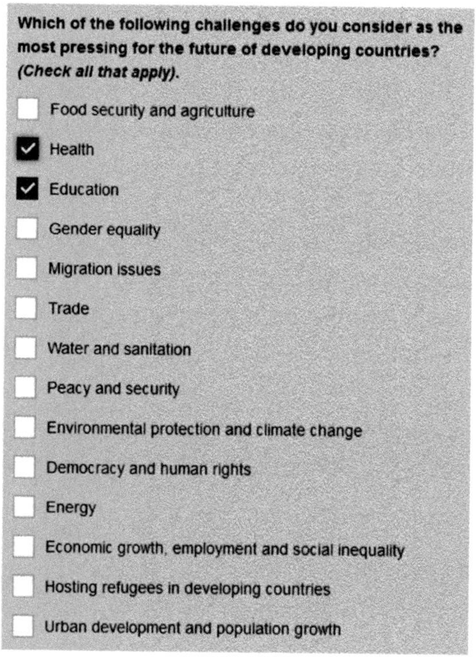

**Which of the following challenges do you consider as the most pressing for the future of developing countries? *(Check all that apply).***

- [ ] Food security and agriculture
- [x] Health
- [x] Education
- [ ] Gender equality
- [ ] Migration issues
- [ ] Trade
- [ ] Water and sanitation
- [ ] Peacy and security
- [ ] Environmental protection and climate change
- [ ] Democracy and human rights
- [ ] Energy
- [ ] Economic growth, employment and social inequality
- [ ] Hosting refugees in developing countries
- [ ] Urban development and population growth

FIGURE 8.11 A closed question with arbitrary answers.

### 8.6.7 Other Aspects

There are some other aspects in which online polls differ from interviewer-assisted polls. One of these aspects is including *sensitive questions*. There are indications that respondents give more *honest* answers to such questions in self-administered polls, such as online polls. The presence of interviewers may very well lead to *socially desirable answers*.

CAPI and CATI questionnaires often apply some form of *routing*. Routing instructions see to it that relevant questions are asked, and irrelevant questions are skipped. So the computer decides the next question to be asked, and not the respondent. Many online polls do not have built-in routing. Respondents are free to jump back and forth through the questionnaire. There is a risk that not all relevant questions will be answered. One may wonder what will produce the best results in terms of data quality and response rates? Should one enforce routing or give complete freedom?

CAI has the advantage that some form of *error checking* can be implemented in the interviewing software, that is, answers to questions are checked for consistency. Errors can be detected during the interview and therefore also corrected during the interview. It has been shown, see for example Couper et al. (1998), that this improves the quality of the collected data. The question is now whether error checking should be implemented in an online poll? What happens when respondents are confronted with error messages? Will they just correct their mistakes, or will they become annoyed and stop answering questions? These issues are discussed in more detail by Couper (2008).

A last aspect to be mentioned here is the length of the questionnaire. If it is too long, respondents may refuse to participate, or they may stop somewhere in the middle of the questionnaire. Questionnaires of CAPI surveys can be longer than those of CATI and online polls. It is more difficult to stop a face-to-face conversation with an interviewer than to hang up the phone or to stop somewhere in the middle of questionnaire of an online poll. Literature seems to suggest that CATI interviews should not last longer than 50 minutes, and completing a questionnaire of an online poll should not take more than 15 minutes.

## 8.7 ONLINE PANELS

Properly setting up and carrying out a poll is often costly and time-consuming. The researcher has to find a sampling frame, select a sample, approach sampled people, persuade them to cooperate, and so on. If an organization is regularly conducting polls, it may consider setting up a *panel*.

A sample of people is invited to become member of the panel. All people in the panel (or a sample) are asked regularly (say, once a month) to complete a questionnaire on the internet.

*Online panels* have become increasingly popular, particularly in the world of market research. This is not surprising, as it is a simple, fast, and inexpensive way of conducting polls. Once an online panel has been put into place, it is simple to conduct a poll. No complex sample selection procedures are required. It is just a matter of sending an e-mail to (a sample of) panel members. No interviewers are involved, and there are no mail costs for sending paper questionnaires. It suffices to put the questionnaire on the internet.

Speed is another advantage of using online panels. A new poll can be launched quickly. There are examples of online polls in which questionnaire design, data collection, analysis, and publication were all carried out on the same day. Online panels have become a powerful tool for opinion polls. For example, in the last weeks of the campaign for the general election of 2012 in The Netherlands, there were four different major national polls each day, and they all used online panels. Moreover, during the campaign for the general election in the United Kingdom in 2015, there were at least seven polling organizations using online panels.

Panels can be used in two different ways. The first one is *longitudinal research*, in which the same set of variables is measured for the same group of individuals at different points of time. The focus of research is on measuring change. The second way to use a panel is *cross-sectional research*. The panel is used as a sampling frame for specific polls that may address different topics and thus measure different variables. Moreover, samples may be selected from specific groups (e.g., elderly, high-educated, or voters for a political party).

This section discusses some issues related to cross-sectional use of online panels. The principles of probability sampling apply: recruitment and sampling must be based on probability sampling. Examples of such panels (see Figure 8.12) are the ELIPSS panel in France (Lynn, 2003), the LISS panel in The Netherlands (Scherpenzeel, 2008), the Knowledge panel in the United States (GfK, 2013), and the GESIS panel in Germany (GESIS, 2014).

It is important to use probability sampling for online panels. To promote proper recruitment techniques, the *Open Probability-based Panel Alliance* was established in 2016. Initiators were the GESIS panel in Germany, the LISS panel in The Netherlands, and the Understanding

FIGURE 8.12   Online panels based on probability sampling.

America Study (UAS) in the United States of America. The alliance facilitates cross-cultural survey research with probability-based internet panels across the globe, endorsing joint methodological standards to yield representative data.

A first issue of online panels is undercoverage. An online panel may suffer from undercoverage, because the target population is usually much wider than just persons with access to the internet. Internet coverage is so high in some countries (the Scandinavian countries, and The Netherlands) that undercoverage is hardly an issue. However, undercoverage is still substantial in many countries. This means that estimates can be biased. One way to reduce coverage problems is to provide internet access to those without it.

Setting up a representative online panel is not simple. Usually, there is no sampling frame of e-mail addresses. So people have to be recruited in some other way. For example, Statistics Netherlands uses the population register to select samples. Selected persons can be approached in different ways: by mail, by telephone, or face-to-face. Face-to-face (CAPI) recruitment is known to produce the highest response rates, but it is also the most expensive mode. CATI is somewhat less costly, but it has the drawback that only those listed in the telephone directory can be contacted. This could be avoided by applying random digit dialing to select the sample, but this has the drawback that there will be no information at all about the nonrespondents. The cheapest mode to recruit persons for an online panel is by sending an invitation letter by ordinary mail. However, this approach is known to have low response rates.

Setting up a representative online panel requires a lot of efforts and costs. This may be the reason that many online panels are not based on the principles of probability sampling, but on self-selection. Respondents are those who happen to have internet, encounter an invitation, visit the appropriate website, and spontaneously decide to become a member of the panel. Unfortunately, such panels lack representativity, and therefore their outcomes may be invalid. For example, during the campaign for the general election in the United Kingdom in 2015, there were seven large polling organizations that used an online web panel: Populus, YouGov, Survation, Panelbase, Opinium, TNS, and BMG. All these panels were formed by means of self-selection. Section 8.8 describes how the use of these panels led to wrong predictions.

Nonresponse is another important problem in online panels. Nonresponse occurs in the following two phases of an online panel: (1) during the recruitment phase, and (2) in the specific polls taken from the panel. *Recruitment nonresponse* may be high, because participating in a panel requires substantial commitment and effort of people. Nonresponse in a specific poll is often low as the invitation to participate in it is a consequence of agreeing to become a panel member. Causes of nonresponse are not at home, not interested in a specific topic, and not able (e.g., due to illness). Nonresponse need not be permanent. After skipping one of the specific polls, a panel member may decide to participate again in a subsequent poll.

Online panels may also suffer from *attrition*. This is a specific type of nonresponse. People get tired of having to complete the questionnaires of the specific polls and decide to stop their cooperation. Once they stop, they will never start again.

The LISS panel is an example of an online panel. It consists of approximately 5000 households. LISS stands for *Longitudinal Internet Studies for the Social Sciences*. The panel was set up by CentERdata, a research institute in The Netherlands. Objective of the panel was to provide a laboratory for the development of and testing of new, innovative research techniques.

The panel is based on a true probability sample of households drawn from the population register in The Netherlands. Telephone numbers were added to selected names and addresses. This was only possible for households with listed numbers. Such households were contacted by means of CATI. Addresses that could not be contacted by telephone were visited by interviewers (CAPI).

TABLE 8.3   Response Rates in the LISS Panel

| Phase | Response (%) |
|---|---|
| Recruitment contact | 91 |
| Recruitment interview | 75 |
| Agree to participate in panel | 54 |
| Active in panel in 2007 | 48 |
| Active in panel in 2008 | 41 |
| Active in panel in 2009 | 36 |
| Active in panel in 2010 | 33 |

Households without internet access and those who worried that filling in a questionnaire on the internet would be too complicated for them were offered a simple-to-operate computer with internet access. This computer could be installed and used for the duration of the panel. This reduced undercoverage problems.

Table 8.3 shows the response rates during the recruitment and the use of the LISS panel. The data are taken from Scherpenzeel and Schouten (2011).

Of the households in the initial sample, 91% could be contacted. So the noncontact rate was 9%. In 75% of the cases, it was possible to conduct a recruitment interview. Of the people in the original sample, only 54% agreed to become a member of the panel. So the recruitment rate was 54%. Over the years, the percentage of respondents dropped to 33%. This is the effect of attrition. So, one out of three persons was still active in 2010.

There will always be nonresponse in an online panel, both in the recruitment phase and in the specific polls. To avoid drawing wrong conclusions, some kind of correction must be carried out. Usually, adjustment weighting is applied. A vital ingredient of weighting is the availability of a set of proper auxiliary variables. These variables must have been measured in the panel, and moreover their population distribution must be known. Weighting adjustment is only effective if two conditions are satisfied: (1) the auxiliary variables must be correlated with response behavior, and (2) the auxiliary variables must be correlated with the target variables.

It is wise to conduct weighting in two steps. First, a weighting technique is applied to correct for recruitment nonresponse. Second, another weighting technique is applied to correct for nonresponse in a specific poll taken from the panel. These weighting techniques differ, because they require different variables. Adjustment weighting for recruitment nonresponse is often difficult, because the number of available auxiliary variables is limited.

Adjustment weighting for a specific poll is more promising, because there can be a lot more auxiliary variables available. Typically, all members of an online panel complete a so-called profile poll when they are recruited. This profile information is recorded for all panel members. Therefore, all profile variables can be used for weighting adjustment of a specific poll. It is important to realize in the design stage of an online panel that profile information may be needed for nonresponse correction. This may help us to select the proper variables for the profile poll.

It is important to keep the composition of the panel stable over time. Only then can changes over time be attributed to real changes in society, and not to changes in the panel composition. An online panel may become less representative due to attrition. This makes it important to carry out *panel refreshment* at certain times. The question is how to do this properly? At first sight, one could think of adding a fresh random sample from the population to the web panel. However, this does not improve the representativity. Those with the highest attrition probabilities remain underrepresented.

Ideally, the refreshment sample should be selected in such a way that the new members resemble the members that have disappeared due to attrition. The refreshment sample should focus on getting people from these groups in the panel. It should be realized that due to refreshment not all members in the panel will have the same selection probabilities. The researcher should take this into account when computing unbiased estimates.

Being in a panel for a long time may have an effect on the behavior and attitudes of the panel members, and even be the cause of a bias. For example, persons may learn how to follow the shortest route through a questionnaire. This effect is called *panel conditioning*. Panel conditioning may be avoided by restricting panel membership to a specific time period. The maximum time period depends on frequency of the specific polls, the length of the poll questionnaires, and also on the variation in poll topics.

## 8.8 EXAMPLE: THE UK POLLING DISASTER

Some called it a polling disaster. All preelection polls for the general election in the United Kingdom on May 7, 2015 were wrong with their prediction of the final election result, and they were consistently wrong. All polls predicted a neck-and-neck race between the Conservative Party and the Labour Party, likely leading Britain to a *hung parliament*. The real election result was completely different. The Conservative Party got much more votes than the Labour Party. The difference was 6.5%.

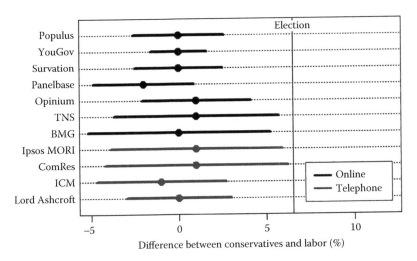

FIGURE 8.13   The polling disaster in the United Kingdom on May 7, 2015.

The dot plot in Figure 8.13 summarizes the polling disaster. It compares the predictions of the difference between the Conservative Party and the Labour Party with the true election result. The big dots represent the predictions of the polling companies just before the election. It is clear that all differences are around zero. The vertical line represents the election result, which is 6.5%. So, the conservatives got 6.5 percentage points more than labor.

The horizontal line segments represent the margins of error. Note that the election result is outside these margins. This means that all poll results differ significantly from the value to be estimated. So, all polls are wrong.

The poll disaster caused a lot of concern about the reliability of the election polls carried out in the United Kingdom. Therefore, the British Polling Council, an association of polling organizations that publish polls, decided to set up an independent enquiry to look into the possible causes of the problems and to make recommendations for future polling. The Inquiry panel concluded that the primary cause of the failure was the lack of representativity of the samples selected by the polling companies. Labor voters were systematically overrepresented, and conservative voters were systematically underrepresented. Moreover, weighting adjustment was not effective for reducing the bias in the predictions. See Sturgis et al. (2016) for more details.

The Inquiry panel ruled out other possible causes. One possible cause was differential misreporting of voting intentions. This phenomenon is sometimes also called the *Shy Tory Factor*. It is the effect that more conservative voters than labor voters said in the polls they were not going to vote. This effect would lead to an underrepresentation of conservatives in the polls. The inquiry panel concluded that there was no substantial *Shy Tory Factor*.

There was also no evidence of a so-called *late swing*. This is the phenomenon that people change their mind between the final polls and the day of the election. This would mean that ultimately people voted for a different party than indicated in the polls.

It is remarkable that the polls were consistently wrong. They all had more or less the same bias. This suggests *herding* as a possible cause of the problems. Herding means that polling organizations make design decisions that cause their polls to be closer to the results of other polling organizations. Consequently, they avoid the risk of being the only polling organization with a wrong result. The Inquiry panel could not completely rule out herding.

Figure 8.13 shows that two different modes of data collection were used in the polls. The first seven polling organizations did online polls. They all have online panels, and people in these panels were all recruited by means of self-selection. So, there is no guarantee these panels are representative. This also means that random samples from these online panels lack representativity. Of course, the polling organizations tried to remove or reduce the bias in their estimators by carrying out some kind of weighting adjustment. Unfortunately, the investigation of the Inquiry panel showed they did not succeed in this.

Four of the polling organizations in Figure 8.12 used telephone polls. This mode of data collection is not without problems. The telephone directory cannot be used as a sampling frame. It is far from complete. Many landline telephone numbers are not in this directory, and it contains almost no mobile telephone numbers. So there is a substantial undercoverage problem. The way out for these polling organizations is to use random digit dialing. They generate and call randomly generated telephone numbers (both landline and mobile). Unfortunately, response rates are very low. They often do not exceed 20% and are sometimes even below 10%. This means these polls can have a large nonresponse bias. Therefore, one can conclude that telephone polls are not an alternative for self-selection online polls.

## 8.9 SUMMARY

When looking at the results of a poll, it is not always clear how the data was collected. Was it a face-to-face poll, a telephone poll, a mail poll, or an online poll? It matters which mode of data collection was used. Typically, face-to-face polls are high quality polls. All other types of polls have their problems.

Online polls have become more and more popular. The reason is that, at first sight, online polls offer an easy, fast, and cheap way of collecting a large amount of data. Online polls, however, also have problems. These problems may easily lead to biased outcomes.

The questionnaire of an online poll cannot be completed by people without internet. So the poll misses part of the population. In fact, the population is reduced to only those in the population with internet.

It is not so easy to draw a random sample for an online poll. Therefore, many researchers rely on self-selection for recruiting respondents. Self-selection is a bad way of data collection that often leads to wrong poll outcomes.

Even if there is no undercoverage, and a random sample is selected, there still is nonresponse. This phenomenon can also cause estimates to be biased. Unfortunately, response rates are low in online polls, and this increases the bias of estimates.

Undercoverage, self-selection, and nonresponse all affect the representativity of the polls. It is therefore important to carry out some kind of adjustment-weighting technique. It is unlikely that one can trust the outcomes of an online poll if the response was not weighted, and even if the results are weighted, there is still no guarantee that all problems are solved. So be careful with online polls!

# CHAPTER 9

# Election Polls

## 9.1 VOTING AND POLLING

Many polls are election polls. Such polls focus on measuring voting intention and voting behavior. They attempt to measure the likelihood that people are going to vote. If they intend to vote, they are asked for their election preferences. Election polls can be used for all types of elections. Here, the focus is on polls for three types of elections:

- *A general election*: This is an election in which people vote for their political party of choice. A typical example is the general election in the United Kingdom on May 7, 2015. The main political parties were the Conservative Party, the Labour Party, the UK Independence Party, the Liberal Democrats, the Scottish National Party, and the Green Party.

- *A presidential election*: This is an election in which people vote for the new leader of the country. So, they select a candidate from a list with at least two candidates. A typical example is the presidential election in the United States that took place on November 8, 2016. The two main candidates were Hillary Clinton (Democrat) and Donald Trump (Republican). There were also a number of third party and independent presidential candidates. Some of them only participated in a few states.

- *A referendum*: This is a vote in which all people in a country or an area are asked to give their opinion about or decide on an important political or social question. A typical example is the referendum in the United Kingdom that was held on June 23, 2016 (the *Brexit* referendum). A referendum almost always has just one question. In the case of the Brexit referendum, the question was: *Should the United Kingdom remain a member of the European Union or leave the European Union?* See Figure 9.1 for the ballot paper.

Election polls attempt to predict the outcome of elections. They also help journalists and citizens to understand what is going in election campaigns, which issues are important, how the behavior of candidates affects the result, and how much support there is for specific political changes. These polls usually attract a lot of attention in the media. Particularly during election campaigns, there can be many polls.

Sometimes election polls contradict each other. If a difference is larger than the margin of error, there must be something wrong with at least one of the polls. Figure 9.2 shows an example. Three polls were conducted in The Netherlands at the same time (early March 2016). The polling companies were I&O Research, Ipsos (Politieke Barometer), and Maurice de Hond (Peil.nl). It is clear that there are substantial differences between the outcomes of the three polls. For example, I&O Research predicted 19 seats for the social-democrats, whereas the prediction of Peil.nl was only 8 seats, which is less than half. And Peil.nl predicted 16 seats for Green Party, which is twice as much as the 8 seats predicted by the Politieke Barometer.

There are several significant differences in Figure 9.2. Taking into account that these polls should measure and estimate the same things, the

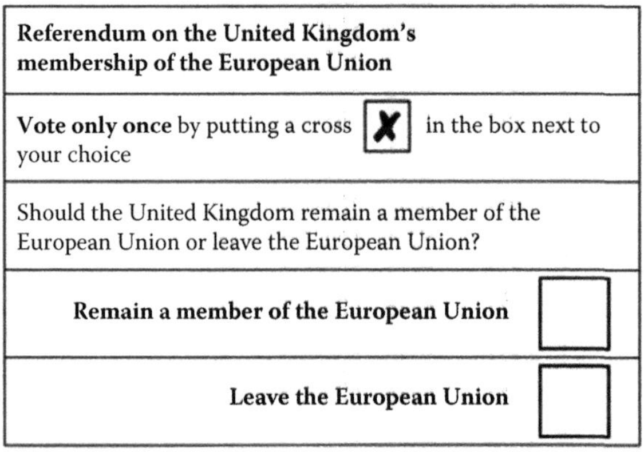

**Referendum on the United Kingdom's membership of the European Union**

**Vote only once** by putting a cross [X] in the box next to your choice

Should the United Kingdom remain a member of the European Union or leave the European Union?

**Remain a member of the European Union** ☐

**Leave the European Union** ☐

FIGURE 9.1 The ballot paper of the Brexit referendum.

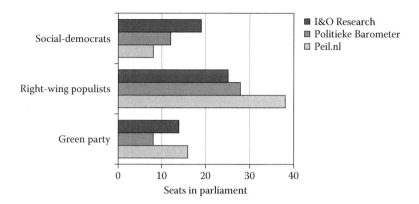

FIGURE 9.2   Three Dutch election polls conducted at the same date.

question arises why they are so different. There can be several causes. Here are a number of them:

- Different target populations were investigated, caused by using different sampling frames.

- The samples were not proper random samples, because they were drawn from different self-selection web panels.

- The text of the questions was different.

- The context in which the questions were asked was different.

- The polls had different response rates.

- Different weighting procedures were used to correct for the lack of representativity.

Unfortunately, poll reports in the media contain too little information about the methodology of these polls. It would be better to have more information that allows readers or viewers to judge the quality of the poll. Chapter 11 (Publication) of this book describes what should be in such a poll report.

This section describes two types of election polls. The first type is the *preelection poll*. Such a poll is conducted before the election takes place. A complication is that the sample consists of people that will vote, that will not vote, and that do not yet know whether they are going to vote. This makes it difficult to compute estimates. The second type of poll is the *exit poll*. Such a poll takes place on the day of the election. The sample is selected in two stages. First, a sample of voting locations is selected. Next, a sample of voters is drawn from each selected voting location. The target population therefore consists of people who just voted.

## 9.2 PREELECTION POLLS

Preelection polls are conducted in the period prior to the election. Such a poll provides insight in the possible turnout of the election, and in the preferences of the voters. The target population consists of all potential voters. Some countries, like the Scandinavian countries and The Netherlands, have a population register. By definition, all people in this register from a certain age (usually 18 years) are potential voters. So, there is no separate voting register, and people do not have to register for voting. The situation is different in countries without a population register, like the United States and the United Kingdom. People can only vote in these countries after they have registered themselves. Ideally, the sample for a preelection poll is selected from the population register or the voting register.

### 9.2.1 Asking for Voting Intentions

The results of an election are determined by the voters. Nonvoters play no role. Therefore, it is important to know which people in the poll are going to vote, and which people are not. Because this is unknown, the election poll has to ask for it. Only those indicating that they will vote are asked the central question of the preelection survey. This is, of course, the question about voting intentions. For which party, candidate, or referendum option will respondents vote?

These two questions (are you going to vote, and for what/whom will you vote?) are the core questions of each preelection poll. There are, however, large differences in the way these questions are asked. Some formats are described in the following.

Probably, the most detailed way in which the two core questions were asked was the approach of the 2016 Daybreak Poll of the *Los Angeles Times*. This online poll was conducted regularly during the campaign of the presidential elections in the United States in 2016. Figure 9.3 shows how the questions could be asked.

A probabilistic approach had been implemented. First, people were asked to specify the chance they would vote. This is a number between 0 (they certainly will not) and 100 (they certainly will). Everything in between expresses at least some doubt. A value of 50 means that voting and not voting are equally probable.

**What is the percent chance that you will vote in the Presidential election?**

[    ] %

**If you do vote in the election, what is the percent chance that you will vote for Trump? And for Clinton? And for someone else?**

Please provide percent chances in the table below

| | | |
|---|---|---|
| Donald Trump (Republican) | [    ] | % |
| Hillary Clinton (Democrat) | [    ] | % |
| Someone else | [    ] | % |
| Total | [    ] | % |

FIGURE 9.3  Asking for voting intentions by asking for chances. (From USC 2016.)

The second question asked the respondents to assign percentages to the various candidates. This was a little more complicated, as the percentages had to add up to 100%. So, this required some basic mathematical skills.

To compute estimates, each respondent was assigned a weight that was proportional to his voting chance. So, respondents with a large chance contributed more than respondents with a small chance. Respondents with voting chance 0 were ignored.

On the one hand, the approach of the *Los Angeles Times* is attractive because most people are familiar with the concepts of chance and percentages. On the other hand, one may wonder whether chances can really be measured at this level of detail. Can people really say whether their voting chance is 64 or 68%?

A somewhat less detailed approach of asking the two core questions was used in an online poll by the polling company BMG during the campaign for the general elections in the United Kingdom in 2015. Respondents were asked how likely they would be to vote in the next election on a scale of 0–10, where 0 meant that they certainly would not vote, and 10 meant that they would certainly vote. The likelihood of them voting was then used to compute weights for respondents. Respondents who answered 10 were assigned a weight 1.0, those answering 9 got a weight 0.9, and so on. Respondents who answered 0 were excluded from the computations (because they will not vote).

All respondents in the BMG poll likely to vote were asked the voting intention question. They could first choose between *Labour, Conservative,*

*Liberal Democrat*, and *UK Independence Party*. These four parties were presented in random order (to prevent response order effects). For respondents in Scotland and Wales, *Scottish National Party*, and *Plaid Cymru*, respectively, were included in the list. There was also a category *Other party*. If this answer was selected, a list of other parties was displayed on the screen.

There are polling organizations in the United Kingdom that used the voting probability question in a different way. They used this question as a filter, and not as a means to compute weights. The idea was to ignore respondents with an answer below a certain threshold. Sometimes the threshold was 10. This means that respondents with an answer in the range from 0 to 9 were ignored. This comes down to the assumption that all respondents with an answer 10 will certainly vote, and all those with an answer 0–9 will not bother to vote. This will probably not be true in practice. There always will be some people who answered 10 and will not vote. And, there will be some people who answered 0–9 and will vote.

The above-mentioned examples attempted to establish voting probabilities by asking for these probabilities. So the question was a numerical one. The voting probability question can also be asked as a closed question. Figure 9.4 contains an example.

The question in Figure 9.4 has five possible answers. This is an odd number with a neutral category in the middle. On the one hand, it is good to have such a middle category, because some people do not know yet whether they will vote or not. On the other hand, the neutral middle category is an easy way out for those not wanting to think about a correct answer.

On the one hand, polling companies sometimes do not include a middle category, because they want to force respondents to give a *real* answer. On the other hand, there often is a separate category *do not know* at the end

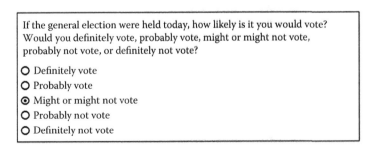

FIGURE 9.4  Asking for voting intention with a rating scale question.

```
If the general election were held today, would you vote?
O  Yes
O  No
◉  Do not know
```

FIGURE 9.5    Asking for voting intention with a yes/no question.

of the list of possible answers. And sometimes, the answer *refused* is also offered, as a separate category, or combined with *do not know.*

Those indicating that they will definitely vote are asked for their voting intentions. Sometimes also probable voters are included. This means that the definite voters and the probable voters care combined into a new group of *likely voters.*

The number of possible answers can reduced even more. See Figure 9.5 for an example. The voting probability question has been reduced to a yes/no question. The element of chance has disappeared. Which answer should someone select who is almost, but not completely, certain about going to vote? *Yes* or *Don't know*? The researcher also runs a risk that many respondents will take the easy way out by choosing the *do not know* option.

It will be clear that only the voters (respondents having answered *Yes*) will be asked to answer the voting intention questions. A question like the one in Figure 9.5 is may be not the best one to measure voting probability. It is too simple. There should be more to choose from.

Van Holsteijn and Veen (2012) have implemented a completely different approach for a preelection poll. They give everybody respondent five votes instead of one. It is up to the respondents what they do with these votes. If they are completely certain about a specific party, they can give all five votes to this party. If they hesitate between two parties, they could give three votes to one party, and two votes to another party. Or they could give four votes to one party and one to the other party. In case of more doubts, they could even decide to give their votes to five different parties. This approach gives a good idea of how stable the support for the parties is. It becomes clear how volatile voters are and whether voters for a certain party also have other parties in mind.

Figure 9.6 shows an example of this approach. It was used in de Dutch preelection poll called *De Stemming.* The fieldwork is carried out by GfK. The voters are divided into three groups: voters giving five votes to a party, voters giving three or four voters to a party, and voters only giving one or two votes to a party.

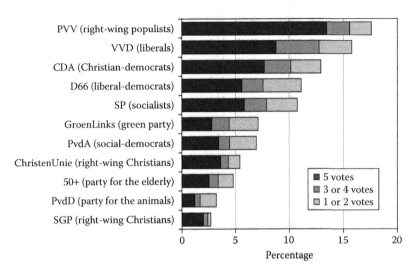

FIGURE 9.6   Showing certain votes, likely votes, and unlikely votes.

The black bars in Figure 9.6 denote the percentages of votes that come from voters giving five votes to the party. These votes can be seen as certain votes. The dark gray bars denote the percentage of votes coming from voters giving three and four votes. These votes can be seen as likely votes. The light gray bars come from voter giving only one or two votes. These votes can be seen as unlikely votes.

The graph shows that particularly the SGP (Staatkundig Gereformeerde Partij, right-wing Christians) and the PVV (Partij voor de Vrijheid, right-wing populists) have stable supporters. Support is a lot less stable for the PvdD (Partij voor de Dieren, party for the animals) and GroenLinks (green party).

To make a prediction for the election, the most likely party is determined for each person in the poll. This is done by selecting the winner in the voting combinations 5, 4-1, 3-2, 3-1-1, and 2-1-1-1. Experiments have shown that the predictions obtained this way, resemble the estimates based on the normal approach.

### 9.2.2  Data Collection

Generally, for the mode of data collection of a preelection poll, a choice is made between a face-to-face poll (CAPI), a telephone poll (CATI), or an online poll. The advantages and disadvantages of these types of polls are discussed in this section.

The first mode of data collection to be discussed here is the face-to-face poll. Usually, the computer-assisted version is used: CAPI. A face-to-face poll means that interviewers visit the people selected in the sample. Interviews take place in the homes of these people. The questionnaire is in electronic form in the computer or tablet of the interviewer. The interview software guides the interviewer and the respondent through the interview.

A strong point of CAPI is that there are interviewers present. They can persuade reluctant persons to participate in the poll, and thus, they reduce nonresponse. They can also assist respondents, if needed, in answering the questions correctly. The survey methodology literature shows that CAPI polls tend to produce high-quality data. But this quality comes at a price. A CAPI survey is time consuming and expensive. There are trained interviewers who have to be paid. They have to travel from one respondent to the next. This also costs time and money.

A weak point of a CAPI poll is that interviewer-assisted polls perform less well with respect to sensitive questions. If there is an interviewer present, respondents tend to give more socially desirable answers to sensitive questions. If voting behavior is considered a sensitive issue, a voting probability question, or a voting intention question, may be answered incorrectly.

To conduct a preelection poll with CAPI, a sample of addresses is required. This means that there must be a sampling frame containing addresses. The ideal sampling frame would be a population register, as this is an up-to-date and complete list. Unfortunately, not every country has such a register. And if such a register exits, there may be restrictions on its use as a sampling frame. Other possible sampling frames are address files, like a postal address file (PAF), or a land registry file.

The use of CAPI for preelection polls is illustrated with an example from Ireland. During the campaign for the general election on February 26, 2016, there were four large polling companies active. Three of them used CAPI (Behaviour and Attitudes, Millward Brown, and Ipsos MRBI). The fourth one (Red C) conducted CATI polls. The sample size of all polls was a little over 1000 persons. This is not surprising as the Association of Irish Market Research Organisations has the guideline that "the minimum sample size for a national published opinion poll should be 1000 respondents."

Typically, polling companies selected samples for their CAPI polls in two steps. Ireland is divided in 40 constituencies. The constituencies are divided into 3440 smaller administrative units (called electoral divisions).

The first step of the sample consisted of selecting a random sample of 100 administrative units. The selected units are called *sampling points*. In each sampling point, 10 people had to be interviewed. This results in a sample of 100 × 10 = 1000 people. There are various ways to select people in a sampling point. One way is called *random route*. Starting at a randomly selected address, the interviewer walks a predetermined route, thereby calling at, say, each fifth house. This can be seen as a form of systematic sampling. It is also a form of quota sampling. The interviewer is instructed to fill quota based on, for example, gender, age, and social class.

A problem with this two-step sampling approach is that respondents are clustered in groups of 10 within administrative units. This may lead to a so-called cluster effect. Respondents within clusters resemble each other. Therefore, interviewing 10 people does not produce much more information than interviewing just one person. This causes estimates to be less precise than one would expect for his sample size. To limit cluster effects, the Association of Irish Market Research Organisations has the guideline that for a sample of 1000 people, no more than 20 people per administrative units may be interviewed. To say it differently: There must be at least 50 different administrative units in the sample.

The second mode of data collection to be discussed here is a telephone poll. The computer-assisted version is almost always used: CATI. A telephone poll means that interviewers call people selected in the sample. Interviews take place by telephone. The questionnaire is in electronic form in the computer of the interviewer. The interview software guides the interviewer and the respondent through the interview.

Like a face-to-face poll, a telephone poll is interviewer assisted. So, interviewers can help to persuade reluctant persons to participate in the poll. They can also assist them with answering questions. The interviewers are not visible. This helps to reduce the risk of socially desirable answers to sensitive questions.

It is not easy to select a simple random sample of telephone numbers. The ideal solution would be to have a complete telephone directory with all landline and mobile telephone numbers. Unfortunately, such a directory does not exist in most countries. Of course, there are telephone directories, but they contain only part of the landline numbers. There are many people with unlisted numbers. Moreover, telephone directories do not contain mobile telephone numbers. So the mobile-only population is completely missing. Unfortunately, the mobile-only group differs with respect to a number of aspects from the rest of the population. For example, they are, on average, younger. This all creates a serious undercoverage problem.

If there is no proper sampling frame available, the researcher may consider random digit dialing (RDD). This comes down to using a computer algorithm for computing random valid telephone numbers (both landline and mobile numbers). One simple way of doing this is to replace the final digit of an existing number by a random different digit. Such an algorithm is able to generate both listed and unlisted numbers. So, there is complete coverage of all people with a telephone. Random digital dialing also has drawbacks. In some countries, it is not clear what an unanswered number means. It can mean that the number is not in use. This is a case of overcoverage, for which no follow-up is needed. It can also mean that someone simply does not answer the phone, which is a case of nonresponse that has to be followed up.

There are polling companies in the United States that use robopolls. A *robopoll* is a telephone poll in which there are no real interviewers. Instead, the computer is completely in control and runs an automated script. Respondents are instructed to answer questions by pressing the proper key on the keypad of their telephone. An important advantage of a robopoll is that it is a lot cheaper than a traditional telephone poll. This is because no (real) interviewers are required. An important disadvantage of robopolls is that there is a federal law prohibiting its use for mobile phones. This restricts its use to landline telephones only.

Dropping response rates of preelection telephone surveys have become a concern. It has become more and more a challenge to achieve sufficiently high response rates. According to the Pew Research Center (2012), response rates of its polls have fallen dramatically. The response rate of telephone polls dropped from 36% in 1997 to only 9% in 2012. There are similar trends in the United Kingdom. According to polling company YouGov, response rates for telephone polls have been declining in recent years to typically below 10%, and they are even much lower in inner city areas. And the Financial Times (2016) quoted Martin Boon, director of ICM research, who said in June 2016 that 20 years ago, it took no more than 4000 telephone calls to get 2000 interviews. Now, the same 2000 interviews take 30,000 telephone calls.

The third mode of data collection to be discussed here is an online poll. The ideal sampling frame for an online preelection poll is a list of e-mail addresses of all members of the target population. Usually, such a list does not exist. Therefore, a different means must be found for selecting a sample. The solution of many polling companies is to use an online panel.

This is a panel of people who have agreed to regularly participate in polls. Such a panel is often large. A sample for an online poll is obtained by drawing a random sample from the panel.

A critical aspect of this approach is the composition of the online panel. A proper panel should have been recruited by means of a simple random sample. If this is the case, the panel is representative, and a random sample from this panel is also representative. More often than not, online panels are not based on a random sample, but on self-selection. Members of such panels are people who spontaneously decide to participate. This does not lead to a representative panel. Consequently, a random sample from a self-selection panel is not representative. So, one has to be careful with polls based on a sample from an online panel.

It should be noted that online polls do not have interviewers. This means that respondents have to answer questions on their own. No one will attempt to persuade them (face-to-face) to fill in the questionnaire of a poll. And, no one will assist them giving the correct answers to the questions. A possible advantage of an online poll is that respondents are more inclined to give a correct answer to a sensitive question. This means that they more honestly report their voting intentions.

For more information about online data collection, see Chapter 8 (online polls), and more in particular Section 8.7 (online panels).

### 9.2.3 Representativity

To be able to compute accurate estimates for the parties, candidates, or referendum options, a representative sample is required. Unfortunately, representativity of a sample can be affected in different ways. The three main problems are undercoverage, self-selection, and nonresponse.

*Undercoverage* is the phenomenon that not everyone in the target population can be found back in the sampling frame. For example, undercoverage occurs in an online poll because there are people in the target population who do not have internet access. Another example is a telephone poll for which only landline telephone numbers are generated by means of RDD. This typically happens in robopolls. In this case, people with mobile telephone numbers will be missing.

*Self-selection* occurs in preelection online polls when the sample is selected from a self-selection online panel. Members of an online panel are those who happened to encounter it on the internet and spontaneously decided to register for it. So, the panel only contains people who like to take part in polls and may be want to earn some money with it.

*Nonresponse* occurs in preelection surveys when the people in the sample do not fill in the poll questionnaire. Causes of nonresponse are noncontact, refusal, or not-able. Nonresponse can happen in face-to-face polls, telephone polls, and online polls. Note that if an online poll is selected from an online panel that was recruited by means of probability sampling, there are two sources of nonresponse: nonresponse in the recruitment phase, and nonresponse in the poll taken from the panel.

As an example, the lack of representativity is shown in a poll taken from an online panel of the Dutch public TV channel NPO 1. In April 2008, the panel contained approximately 45,000 members. All these members were asked to take part in a poll. The number of respondents was 19,392, which comes down to a response rate of 42.5%. The representativity of this poll was affected in three ways. First, only people with internet could become a member of the panel. So, there is undercoverage. This will not be serious as internet coverage is high in The Netherlands (96% in 2015). Second, the members were recruited by means of self-selection. And third, there was a substantial amount of nonresponse.

Figure 9.7 explores the lack of representativity of the variable measuring voting behavior in the last general elections (in 2006). There is a substantial difference between the distribution of this variable in the poll and the population. The percentage of voters in the poll is 94.8% and in the population it is only 80.2%. Hence, voters are clearly overrepresented. Likewise, nonvoters are underrepresented in the poll.

This problem is caused by the strong correlation between voting and participation in a poll. Voters tend to participate in polls, and nonvoters tend to refuse. This phenomenon is often observed in preelection polls.

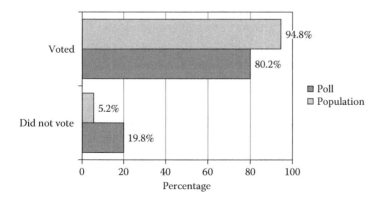

FIGURE 9.7   The distribution of voting behavior in the poll and in the population.

To repair the lack of representativity, a correction must be carried out. Usually, a weighting adjustment technique is applied. This comes down to assigning weights to respondents. Underrepresented group get a weight larger than 1, and overrepresented get a weight smaller than 1. To compute weights, auxiliary variables are required. These variables must have been measured in the poll, and moreover, their population distribution must be available. Often, auxiliary variables like gender, age, and region of the country are used for weighting. However, auxiliary variables are only effective if they satisfy two important conditions:

- They must be correlated with participation behavior.

- They must be correlated with the target variables of the poll (voting behavior).

If these conditions are not satisfied, weighting will have no effect. The bias of estimates will not be removed. The variable that measured voting behavior in the previous election is an important weighting variable. It can be measured in the poll, its population distribution is available (from the Electoral Council), it is correlated with participation behavior (see Figure 9.7), and it is also correlated with current voting behavior.

Unfortunately, it is not always possible to get accurate answers to the question about previous voting behavior. This problem occurs when time has passed between the previous election and the preelection poll. Several effects may lead to wrong answers. An example shows what can go wrong. The members of the panel of the Dutch public TV channel NPO 1 were asked in May 2005 what they voted in the general elections of January 2003. In October 2005 (five months later), the same question was asked again. In Table 9.1, the answers to both questions are compared. The matrix is based on the answers of 15,283 respondents. The columns denote the votes in May 2005, and the rows the votes in October 2005.

Theoretically, the answer given in May should be the same as the answer in October. Therefore, all cells on the diagonal of the matrix should be 100%, and all off-diagonal elements should be empty (0%). Unfortunately, this is not the case. The voters for PvdA (Partij van de Arbeid, social-democrats) seem most consistent. A percentage of 91.1 of the voters in October also said in May that they voted for this party. The least consisted is the SGP (Staatkundig Gereformeerde Partij, Conservative Christian). Only 70.6% of these voters also said in May that they voted for this party.

TABLE 9.1    Vote in the General Elections of 2003

| Vote 2 | Vote 1 | | | | | | | | | |
| --- | --- | --- | --- | --- | --- | --- | --- | --- | --- | --- |
| | CDA | PvdA | VVD | D66 | GL | SP | LPF | CU | SGP | Total |
| CDA | 90.7 | 1.6 | 2.8 | 0.4 | 0.2 | 0.6 | 0.8 | 0.7 | 0.0 | 100.0 |
| PvdA | 0.9 | 91.1 | 0.7 | 1.1 | 0.9 | 2.2 | 0.8 | 0.1 | 0.0 | 100.0 |
| VVD | 2.3 | 0.9 | 90.4 | 1.2 | 0.1 | 0.5 | 2.4 | 0.2 | 0.0 | 100.0 |
| D66 | 1.6 | 3.1 | 2.0 | 86.8 | 0.8 | 2.3 | 0.7 | 0.2 | 0.0 | 100.0 |
| GL | 0.2 | 3.7 | 0.4 | 1.2 | 86.5 | 3.7 | 0.2 | 0.4 | 0.1 | 100.0 |
| SP | 0.8 | 3.9 | 0.3 | 0.7 | 2.0 | 88.6 | 1.0 | 0.3 | 0.2 | 100.0 |
| LPF | 2.2 | 2.5 | 8.5 | 0.8 | 0.2 | 2.1 | 76.9 | 0.3 | 0.1 | 100.0 |
| CU | 3.2 | 0.7 | 1.1 | 0.0 | 0.0 | 0.4 | 0.7 | 91.0 | 0.4 | 100.0 |
| SGP | 0.0 | 2.9 | 5.9 | 2.9 | 0.0 | 0.0 | 2.9 | 8.8 | 70.6 | 100.0 |

*Source:*  EénVandaag Opiniepanel (2005).
*Note:*  Vote 1: asked in May 2005, Vote 2: asked in October 2005. Percentages of row totals.

Note that a substantial amount (8.8%) of October voters said in May that they voted for Christen Unie (CU), also a Christian party. It is clear from this table that the answers to this voting question are not very stable.

Van Elsas et al. (2014) explored voting recall problems in more detail. They found various possible causes. A first cause is that the voting question is a recall question. It asks about events that took place some time ago. The previous election could even be four or five years ago. People make errors when they have to recall events from the past. The longer ago an event took place, the more serious the *recall error*. Recalling past voting is even more difficult if there were other elections in the meantime. For example, between two general elections in The Netherlands, there can be local elections, provincial elections, and European elections. More volatile voters may be confused and therefore report the wrong voting behavior.

A second possible cause is *cognitive bias*. As time passes, current attitudes are gradually replacing past attitudes. In order to show consistent behavior, respondents change past voting behavior.

A third possible bias is *nonattitudes*. Preferences of voters may not be firmly based in strong attitudes about relevant issues. This leads to more superficial, more random, behavior. This makes it more difficult to recall past voting behavior. If respondents have a strong attitudes, it easier for them to recall previous party choice.

The conclusion is that voting at the previous election is an important variable for weighting adjustment, but only if the answers to his question are not subject to too much measurement error. Therefore, it is wise not to use it if the question is asked a number of years after the election. If a polling company maintains on online panel, it is good idea to ask the voting question straight after the day of the election and store it in the panel for later use. This way, the question is almost not subject to recall errors.

### 9.2.4 Single-Question Polls

The internet has made it very easy to conduct single-question polls. Such polls with only one question can often be found on the front page of news media websites. There is just one closed question. Everybody encountering this poll can take part in it. And it only takes a few seconds to complete it. After having answered the question, it is usually replaced by the frequency distribution of the answers (thus far).

It is even more easy to create a single-question poll in *Twitter*. In a few minutes, a tweet can be created containing a closed question. The text of the question may not exceed 140 characters. There can be up to four answer options, and the text of each answer option may not exceed 25 characters. By default, a Twitter poll is active for one day, but this may be extended to seven days.

Single-question polls have serious defects. The most important problem is the lack of representativity. All single-question polls are based on self-selection of respondents. The question is only answered by people who visit the website and spontaneously decide to participate in this poll. Usually, the group of visitors is far from representative.

Another big disadvantage of a single-question poll is that it is not possible to apply adjustment weighting to repair the lack of representativity. Weighting adjustment requires auxiliary variables that must have been measured in the poll. These auxiliary variables are missing in a single-question poll.

There is also a risk that a single-question poll is manipulated. The question can be answered multiple times by using different internet devices (PC, laptop, tablet, or smartphone). Sometimes there is also no protection against answering the question multiple times with the same device.

It is not clear what the target population of a single-question poll is. Everyone can participate. For example, people from Europe could participate in a poll about the presidential elections in the United States.

There is an increasing risk that a single question poll is manipulated by a votebot. A votebot is a special kind of internet bot. An *internet bot* is a software application that carries out automated tasks (scripts) over

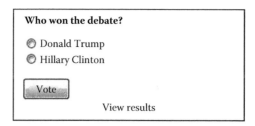

FIGURE 9.8 A single question similar to that of the Breitbart poll. (From Breitbart 2016.)

the internet. A *votebot* aims to vote automatically in online polls, often in a malicious manner. A votebot attempts to act like a human but conducts voting in an automated manner in order to manipulate the result of the poll. Votebots are sold on the internet, but simple votebots are easy to code and deploy. Developers of online polling applications can protect their software against attacks of votebots, but this requires extra efforts.

Many single-question pols were conducted during the campaign for the presidential election on November 8, 2016 in the United States. They were particularly popular for measuring the performance of the candidates in the debates. There were three debates between Hillary Clinton and Donald Trump. After the third debate, there was a single-question poll on the website of Breitbart News Network. This is a politically conservative American news, opinion, and propaganda website. The question in Figure 9.8 is similar to the one of Breitbart. Apparently, there are only two choices. There is no option for those who do not know who won the debate.

The poll was activated immediately after the end of the debate. Since Breitbart is a conservative news network, many visitors will be Republicans. Therefore, it was to be expected that Republican Donald Trump would get more votes than the Democrat Hillary Clinton. The left-hand side of Figure 9.9 shows the results of the poll after a couple of hours. At that moment, over 150,000 people had answered the poll question. Surprisingly, Hillary Clinton was in the lead. According to over 60% of the respondents, she had won the debate.

Breitbart claimed that the poll was infiltrated by votebots that were operated from countries such as Romania, Germany, and South Korea. Based on these votes, Hillary Clinton was overwhelmingly the winner of the final presidential debate. Later, after 300,000 people voted, the situation changed, and Donald Trump took the lead with 54% of the votes, see Figure 9.9.

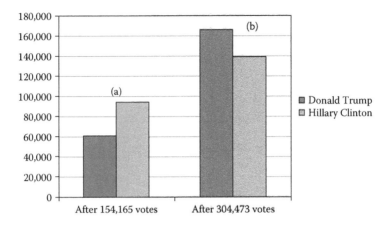

FIGURE 9.9 The results of the Breitbart poll after approximately 150,000 respondents (a) and approximately 300,000 respondents (b).

Probably, there were also votebots working for Donald Trump. This example clearly shows that representativity of this single-question poll is seriously affected by self-selection, manipulation by votebots, and ignoring weighting adjustment. Therefore, the outcomes of this poll should not be trusted.

### 9.2.5 An Example: The U.S. Presidential Election in 2016

It was made clear in this section that it is not so easy to carry out a pre-election poll which produces accurate estimates. There are several critical factors that may have a negative effect on the outcomes of such a poll. A well-known example of an event where polls did not perform well was the general elections in the United Kingdom on May 7, 2015. All polls predicted a neck-and-neck race between The Conservative Party and The Labour Party. The election resulted, however, in a comfortably majority for the Conservative Party. The polls were systematically wrong. Some called this the *UK polling disaster*. See Section 8.8 for more details.

Polls had also problems predicting the outcomes of the presidential election in the United States on November 8, 2016. Almost all polls predicted that Hillary Clinton would be the new president, but the surprising result was that Donald Trump was elected.

Table 9.2 contains an overview of 12 polls that were carried out one or two days before the election. The data come from Wikipedia (2016). In all these polls, respondents could choose between four candidates, of which Hillary Clinton and Donald Trump were by far the most important ones.

TABLE 9.2   Predictions of Polls One or Two Days before the U.S. Presidential Election

| Peiler | Datum | Sample | Mode | Clinton (%) | Trump (%) | Difference |
|--------|-------|--------|------|-------------|-----------|------------|
| YouGov | November 4–7 | 3677 | Online | 45 | 41 | 4 |
| Insights West | November 4–7 | 940 | Online | 49 | 45 | 4 |
| Bloomberg | November 4–6 | 799 | Phone | 44 | 41 | 3 |
| Gravis | November 3–6 | 16,639 | Robopoll | 47 | 43 | 4 |
| ABC News | November 3–6 | 2220 | Phone | 47 | 43 | 4 |
| Fox News | November 3–6 | 1295 | Phone | 48 | 44 | 4 |
| IBD/TIPP | November 3–6 | 1026 | Phone | 41 | 43 | −2 |
| Monmouth | November 3–6 | 802 | Phone | 50 | 44 | 6 |
| Ipsos | November 2–6 | 2195 | Online | 42 | 39 | 3 |
| CBS News | November 2–6 | 1426 | Phone | 45 | 41 | 4 |
| Rasmussen | November 2–6 | 1500 | Robopoll | 45 | 43 | 2 |
| NBC News | October 31–November 6 | 70,194 | Online | 47 | 41 | 6 |
| Election | November 8 | | | 48.1 | 46.0 | 2.1 |

Source: Wikipedia (2016).

Four of the 12 polls were online polls. Their samples were selected from online panels. Six polls were telephone polls. Samples of telephone numbers (both landline and mobile numbers) were selected by means of RDD.

The table contains two robopolls. A robopoll is a fully automated telephone poll. A computer is in control of sample selection and interviewing. There are no interviewers involved. A federal law in the United States prohibits the use of robopolls of mobile numbers. To avoid undercoverage due to missing mobile numbers, additional data are collected in an online poll.

The poll of NBC News had a very large sample size of 70,194 people. This sample was selected by SurveyMonkey. This is a website on which online polls can be created and carried out. Each day, about three million people fill in questionnaires on the SurveyMonkey platform. The sample has been selected from these users. One may wonder how representative such a sample is. Another outlier is the IBD/TIPP poll. It is to only poll in the table that predicted more votes for Trump than for Clinton.

Figure 9.10 compares the outcomes of the polls with the real election result. The dot plot in Figure 9.10a shows the figures for Clinton and Figure 9.10b does the same for Trump. The black dots represent the predictions of the polls. The horizontal black line segments denote the margins of error of the predictions (assuming simple random sampling). The vertical black line is the true election result.

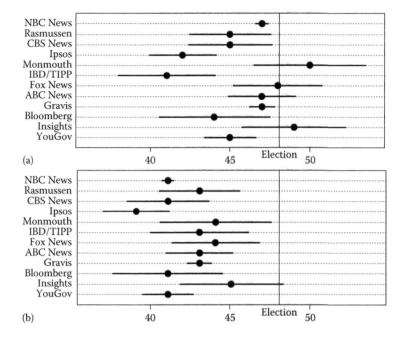

FIGURE 9.10 Polls conducted one or two days before the U.S. presidential election, November 8, 2016: (a) dot plot shows the predictions for Clinton and (b) dot plot shows the predictions for Trump.

In the case of Clinton, several polls are reasonably close to the real election result (48.1%). For four polls, the outcomes are within the margins of errors. For eight polls, there is a significant difference. Also for Trump, the polls do not perform well. Eight of the 12 polls predicted a percentage for Trump that was significantly too low. So, Trump did a lot better than predicted. In summary, one can say that the polls predicted 3% or 4% points more for Clinton, whereas in reality, the difference was 2.1%.

Figure 9.10 also shows that large polls are not necessarily better than small polls. The poll of NBC News is based on a large sample of around 70,000 people. Still, its prediction for Trump is completely wrong. It predicted 41% instead of 48.1%. So, if the poll has a defect (the online sample lacks representativity), it cannot be repaired with a larger sample. What helps is to make the sample more representative.

Why were the polls significantly wrong? Finding an explanation for these problems requires more research. Some possible causes can be suggested by comparing these polls with those of the general election in the United Kingdom on May 7, 2015.

A first possible cause is the lack of representativity. Two types of polls were conducted: online polls (for which samples were selected from a self-selection online panel) and telephone polls (based on RDD, and with very low response rates). The same types of polls were used in the United Kingdom. An Inquiry panel investigated the UK polling disaster. It concluded that the samples were not representative. Moreover, weighting adjustment techniques were not effective. They could not repair the lack of representativity.

A second possible cause could be a *Shy Trump Factor*. Voters do not want to admit in the polls that they are going to vote for Donald Trump. There seem to be some indications for this. In the United Kingdom, a possible *Shy Tory Factor* was investigated. This would occur when people do not want to admit in a poll that they are changing from Labour to Conservative. The Inquiry panel concluded that there was no *Shy Tory Factor*.

A third possible cause is a *Late swing*. This is the phenomenon that people change their mind after the last opinion poll and decide to vote for a different party or candidate. The UK Inquiry panel could not find any evidence of a *Late swing*.

A fourth possible problem is that polls encounter difficulties determining whether people will vote or not. If this decision is based on previous voting behavior, people now voting for Trump and not voting in the past will be excluded from the polls. This could cause support for Trump to be underestimated.

The problems in the U.S. polls will be investigated by the American Association of Public Opinion Research (AAPOR). Hopefully, this leads to better polls in the future.

## 9.3  EXIT POLLS

Section 9.2 described *preelection polls*. Such polls are conducted before the election takes place. A complication of these polls is that the sample will not only contain voters but also nonvoters, and those having not yet made up their mind. Therefore, it is no easy to compute accurate estimates.

This section is devoted to *exit polls*. Such polls are conducted on the day of the election. The sample is selected in voting locations. Respondents fill in the questionnaire in the voting locations, immediately after they voted. Therefore, all respondents are voters. Nonvoters are excluded. Respondents are asked to complete a questionnaire with a few questions. The most important question asked is what the respondents just voted.

Why carry out an exit poll? There are three possible reasons:

- An exit poll can be carried to obtain a very early indication of the outcomes of an election. A well-organized exit poll can produce estimates straight after the voting locations close. Counting (and possibly recounting) the election votes can take a long time. Media do not want to wait for this. They want early results.

- An exit poll need not be a single-question poll. There can be more questions. The poll can also record the vote for a previous election, and possibly some demographic variables, like gender and age. This makes it possibly to analyze voting behavior and, in particular, changes in voting behavior.

- An exit poll can be used as an instrument to check the voting process. If there are large difference between the outcomes of an exit poll in a polling location and its election counts, this could be an indication of election fraud.

### 9.3.1 The First Exit Poll

The first exit poll ever was carried by the Dutch sociologist Marcel van Dam on February 15, 1967 in the town of Utrecht in The Netherlands. There were general elections on this day. He selected four voting locations from the 164 voting locations in the town. All voters in the four locations were asked to fill in a small questionnaire form. This questionnaire is reproduced in Figure 9.11.

The questionnaire contains four questions. The first question asked for which party the respondent just voted. This question is always asked in an exit poll. The second question asks for which party the respondent voted in the previous election. In this case, the previous elections were local elections, which were held on June 1, 1966. There are two important reasons for asking such a question. The first one is that it can be used to explore how people changed their political views between the two elections. The second reason is that this variable can used for weighting adjustment. The third question records the age group of the respondent, and the fourth question asks about religion. Note that there were questionnaires in two colors. Male respondents filled in a white form, and female respondents a yellow one. So also gender was recorded.

WILT U HET HOKJE ACHTER HET
JUISTE ANTWOORD ROOD MAKEN?
(ER ZIJN VIER VRAGEN).

VRAAG 1  Op welke partij hebt U zojuist gestemd?

| K.V.P. | ☐ | Boerenpartij | ☐ |
| P.v.d.A. | ☐ | G.P.V. | ☐ |
| V.V.D. | ☐ | Lib. Volkspartij | ☐ |
| A.R.P. | ☐ | Noodraad | ☐ |
| C.H.U. | ☐ | Landsbelangen | ☐ |
| C.P.N. | ☐ | Machiela | ☐ |
| P.S.P. | ☐ | C.D.U. | ☐ |
| S.G.P. | ☐ | Democr. '66 | ☐ |
| Lijst Voogd | ☐ | Partij v. Ongehuwd. | ☐ |

Blanco (ongeldig)  ☐

VRAAG 2  Op welke partij hebt U de laatste keer, dus bij de gemeenteraads-verkiezingen, 1 juni 1966, gestemd?

Was nog te jong om te stemmen  ☐

| K.V.P. | ☐ | Boerenpartij | ☐ |
| P.v.d.A. | ☐ | G.P.V. | ☐ |
| V.V.D. | ☐ | C.D.U. | ☐ |
| A.R.P. (+S.G.P.) | ☐ | Partij v. Ongehuwd. | ☐ |
| C.H.U. | ☐ | Wereldburgers | ☐ |
| C.P.N. | ☐ | Volksw. en Refer. | ☐ |
| P.S.P. | ☐ | | |

Blanco (ongeldig) }
Niet gestemd     }  ☐

VRAAG 3  In welke leeftijdsgroep hoort U?

| 21 tot en met 30 jaar | ☐ |
| 31 tot en met 40 jaar | ☐ |
| 41 tot en met 50 jaar | ☐ |
| 51 tot en met 65 jaar | ☐ |
| 66 jaar en ouder | ☐ |

VRAAG 4  Tot welk Kerkgenootschap rekent U zich?

| Rooms Katholiek | ☐ |
| Nederlands Hervormd | ☐ |
| Gereform. Gezindten | ☐ |
| Andere | ☐ |
| Geen | ☐ |

WILT U HET FORMULIER INGEVULD
IN DE BUS DOEN?
DANK U!

FIGURE 9.11  Questionnaire for the exit poll in Utrecht, The Netherlands, 1967.

The four voting locations were not selected by means of random sampling. They were purposively selected in such a way that all social classes were presented. Moreover, voting locations with disturbing conditions (like large old people homes, or convents) were avoided.

In each voting location, all voters were invited to complete the exit poll questionnaire. There was ample time for this as the official voting procedure took some time (voter registration, unfolding the ballot paper, voting, folding of the ballot paper, and putting it in the ballot-box). A total of 3408 voters participated in the exit poll. This comes down to a response rate of 85%, which is very high.

Figure 9.12 shows how well the exit poll (based on four voting locations) could predict the results for the town of Utrecht (164 voting locations). The estimates are very close to the real election results. The largest difference is only 0.5% point.

All details of the exit poll in Utrecht are described in Van Dam and Beishuizen (1967). The first exit poll in the United States was conducted by Warren Mitofsky in November 1967. He carried out this exit poll for a governor's election in the U.S. state of Kentucky. See Scheuren and Alvey (2008) for more details.

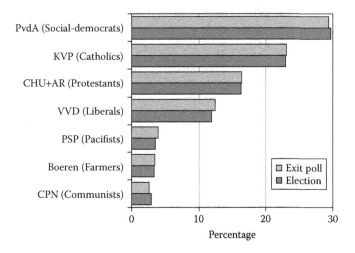

FIGURE 9.12 Results of the exit poll in Utrecht in 1967. (From Van Dam and Beishuizen 1967.)

### 9.3.2 The Basics of Exit Polls

This subsection describes various aspects of exit polls. Starting point is the target population. This is the group of all people who come to the voting locations. All these voters can be asked to participate in the exit poll after they have voted. It should be noted that the target population of an exit poll differs from that of a preelection poll. The target population of a preelection poll consists of those who intend to vote, who intend not to vote, and those who do not know yet.

In principle, all voters can be contacted when they leave the voting location after having voted. There are some exceptions:

- *Voting by mail*: In some elections, it is possible to vote by mail. People can send their ballot paper by mail to the authorities. Usually, this can be done in a period before the day of the election. These people do not show up in a voting location.

- *Early voting*: In some elections, there is the possibility that people can already vote in person in voting locations prior to the day of the election. Some states in the United States offer this facility. These voters are missed if the exit poll is only conducted on election day.

- *Proxy voting.* If voters are unable or unwilling to vote in their allocated voting location, they can decide to authorize someone else to vote on their behalf. It is unclear whether the researchers will detect that a voter has voted more than once.

- *Voting elsewhere.* In some countries, it is possible (after a request) to vote in a different voting location. For example, in The Netherlands, people can vote in a railway station while they are on their way to work. Such people will not show up in their own voting location.

These exceptions cause not all voters to be available for an exit poll. Fortunately, these special groups are usually small so that the outcomes of the exit poll are not seriously affected.

Data collection for an exit poll takes place in or just outside voting locations. After they have voted, respondents are asked to fill in the (paper) exit poll questionnaire. They drop the completed questionnaire in the exit poll ballot box. There are no interviewers. Respondents choose options on their own. This approach reduces the risk that respondents give socially desirable answers or give no answers at all (nonresponse).

The questionnaire of an exit poll can contain only a few questions. There simply is no time to let voters complete a large questionnaire. It would lead to congestion inside or outside the voting location and cause a substantial amount of nonresponse. If exit poll respondents cannot be processed as fast as the voters in the voting location, it may be better to draw a sample of voters at each location. This sample must be representative for all voters at this location. It is also important that sample voters are well-distributed over the opening period of the voting location, as different groups of voters may vote at different times. Therefore, the choice is often for a systematic sample: every $n$th voter is asked to participate in the exit poll. The value of $n$ could, for example, be equal to 5, in which case each fifth voter is included in the poll.

A country usually has a large number of voting locations. For example, in the United Kingdom, there were around 50,000 voting locations for the general elections in 2015. Even in small country as The Netherlands, there were approximately 9000 voting locations for the provincial elections in 2015. There are too many voting locations to include them all in an exit poll. So, a sample of voting locations is selected. This is often not done by means of simple random sampling. Instead, purposive selection is applied in such a way that the sample becomes representative with respect to the results of the previous election, the geographical distribution over the country, and the distribution over rural and urban areas. The sample in the United Kingdom consisted of 141 polling locations, and the sample in The Netherlands had 43 voting locations.

The disadvantage of this sampling approach is that no margins of error can be computed. Only if the purposive sample resembles more or less a random sample, can margins of errors be computed as an approximation.

In its most simple form, the questionnaire of an exit poll has only one question. This question asks what one just voted in the voting location. In principle, this question is sufficient for estimating the results of the voting location. There are also exit poll questionnaires containing a few more questionnaires. An interesting one is the question asking for voting behavior in the previous elections. By comparing the distribution of this question in the exit poll with the official result, one can check whether or not the exit poll is representative. If not, some form of weighting can be carried out. Of course, this variable can also be used for analyzing changes between the current and the previous election. Note that one should be careful with this question. Because it asks about events that may have taken place already some years ago, its answers may not be accurate. One could also ask some demographic questions, like gender and age.

A nice property of exit polls is that their response rates are often high. The first exit poll (in Utrecht, The Netherlands) had a response rate of 85%. The exit poll in 2015 in the United Kingdom had a response rate of 75%. And, the response rate of the exit poll of Ipsos for 2015 elections in The Netherlands was 82%. The high response rates are not surprising, taking into account that response in polls is correlated with voting. So, voters have a higher tendency to participate in a poll than nonvoters. Moreover, it is not so easy to avoid the researchers outside voting locations. And just a few questions have to be answered.

Speed is an important property of an exit poll. As soon as the voting locations close at the end of the day, the results of the exit poll should be available immediately for publication in the media. This may mean that the exit poll stops a little earlier, to be able the process the data and compute estimates. The exit poll should not stop too early, as this may cause a bias in the estimates. It is not unlikely that late voters differ, on average, from early voters.

If exit poll data are capable of making accurate estimates of the election results, one may wonder whether it is possible to publish intermediate results already in the course of the election day, and not at the end of it. For decades, media organizations have agreed not to reveal this information until voting locations close. Journalists keep that information to themselves, campaign teams use their own techniques to find out what is going on, and voters do not get information about how the election is going. The main concern is that publishing early information about the outcomes of the election may influence people that still have to vote. Moreover, some worry that early publication may reduce election turnout.

Countries have different regulations with respect to early publication of exit polls. For example, in Montenegro and Singapore, it is not allowed to conduct exit polls. In Germany and the United Kingdom, early publication of exit polls is prohibited. In the Czech Republic, exit polls are allowed, but not in voting locations. And in a country like The Netherlands, there is no restriction, although the Electoral Council is aware of the possible effects of early publication. Many more countries have no restrictions.

### 9.3.3 Examples of Exit Polls

An example of a recent exit poll is the one conducted for the general elections of May 7, 2015 in the United Kingdom. The exit poll was commissioned by three TV stations: BBC, ITV, and Sky News. The work was done by the two polling companies GfK NOP and Ipsos MORI. Spread over 133 constituencies, 141 voting locations were selected. This was not a random sample. There was a careful purposive selection process. Voting locations were taken from constituencies with different characteristics. Moreover, as much as possible, the same voting locations were used as in the previous elections. A team of around 400 researchers stood outside the voting locations and asked voters to fill in dummy ballot papers, and depositing them into an exit poll box. Not every voter in the voting location was asked to participate in the exit poll. A systematic sample of between 100 and 200 was taken in each voting location (every nth voter). A total sample of the exit poll consisted of around 22,000 voters.

The selected voters were asked to answer only one question, and this was the party for which they just voted. So, there were no questions about voting behavior in the previous election, and various demographic characteristics. Consequently, the collected data could not be used of in-depth analysis.

The response rate of this poll was around 75%, which is a high response rate. Figure 9.13 compares the outcomes of the exit poll with the election result. The estimated numbers of seats in parliament are close to the true

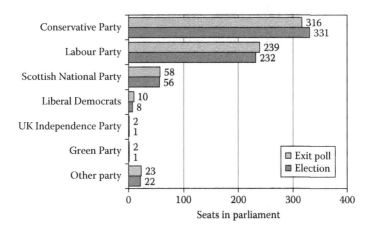

FIGURE 9.13   Results of the exit poll in the United Kingdom in 2015.

numbers of seats. It is clear that there was no neck-and-neck race between the Conservatives and Labour as was predicted by all preelection polls (see Section 8.8 about the UK polling disaster).

A second example of an exit poll is the one conducted for the provincial elections in The Netherlands on March 18, 2015. It was conducted by Ipsos. Out of a total of 9266 voting locations, a purposive sample of 43 locations was selected. The sample was made representative with respect to the result of the previous elections, and with respect to the distribution over rural and urban areas. In each of the 43 voting locations, all voters were asked to fill in the exit poll questionnaire, which contained just one question (for which party did you just vote?).

A total of 30,871 people came to the selected voting locations to vote. A total of 25,163 agreed to participate in the exit poll, which comes down to a response rate of 82%. It is remarkable that the response rate is just as high as that of the first exit poll in 1967 (85%). Figure 9.14 compares the estimates of the exit poll with the election results. The differences are very small. The largest difference is 0.9% points for the PVV.

## 9.4  SUMMARY

Many polls are election polls. Such polls attempt to predict the outcome of elections. They also help journalists and citizens to understand what is going in election campaigns, which issues are important, how the behavior of candidates affects the election results, and how much support there is for

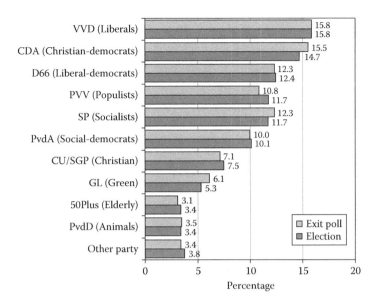

FIGURE 9.14 Results of the exit poll in The Netherlands in 2015.

specific political changes. These polls usually attract a lot of attention in the media. Particularly during election campaigns, there can be many polls.

There are preelection polls and exit polls. A preelection poll is conducted before the election takes place. A complication is that the sample consists of people who will vote, and those who will not vote, and those who do not yet know whether they are going to vote. This makes it difficult to compute estimates.

Exit polls take place on the day of the election. The sample is selected in two stages. First, a sample of voting locations is selected. Next, a sample of voters is drawn from each selected voting locations. The target population therefore consists of people who just voted. Only a few questions can be asked. Response rates are usually high, and estimates are accurate.

CHAPTER **10**

# Analysis

## 10.1 THE ANALYSIS OF POLL DATA

After the researcher has finishing data collection, he has a (hopefully large) number of completed questionnaires. If data were collected with paper questionnaires, the data have to be entered into a computer. If some form of computer-assisted interviewing was used, the data are already in the computer. A next step will be to analyze the data. This is the topic of this chapter.

The first step should always be to check the data for errors. This is sometimes also called *data editing*. Take one variable at the time and look for unusually small or large values, for example, a person with the age of 140 years. These are called *outliers*. An outlier can be a wrong value that must be corrected, but it can also be a correct value. This has to be checked. It is also a good idea to look for unusual combinations of values, like a voter of 12 years. Again, such values can point at an error, or they are unlikely, but correct.

The next step is an *exploratory analysis*. Such an analysis focuses on exploring an (often large) data set. It summarizes the main characteristics of the data. It should reveal the information that is in the data set. Software tools are required to carry out such an analysis. They should help one to discover unexpected patterns and relationships in the data.

Exploratory analysis provides a set of tools and techniques for summarizing large data sets in a few numbers, tables or graphs. If a new, unexpected feature is discovered, the researcher must check whether this really is an interesting new fact, or it is just an artifact of this data set.

An exploratory analysis aims at investigating the collected data, and nothing more. The conclusions only relate to this data set. So no attempt is made to generalize from the response of the poll to the target population. If the data are used for drawing conclusions about the target population as a whole, a different type of analysis should be carried out. This is called an *inductive analysis*. Then, conclusions take the form of estimates of population characteristics. Examples are an estimate of the percentage of people who are going to vote, and an estimate of the mean number of hours people are online. Inductive analysis can also take the form of hypothesis testing. For example, a researcher can test the hypothesis that people are more satisfied with their life than five years ago.

It is typical for inductive analysis that the conclusions are based on sample data and, consequently, have an element of uncertainty. Therefore, one always has to take into account margins of error. Fortunately, the researcher has some control over the margin of error. For example, an increased sample size will reduce the uncertainty.

If the objective is to test a hypothesis about a population, it is better to spilt the data set randomly into two parts. One part can be used for exploration and to formulate a hypothesis. Subsequently, the other part can be used to independently test the hypothesis. It is considered bad practice to use the same data set both for formulating and testing a hypothesis.

When looking at poll results, one has to realize that it is not always simple to compute proper population estimates. The reason is that all kinds of problems can occur during data collection. Therefore, the quality of the data is not always as good as one hopes it to be. Here are a number of possible issues:

- The sample has not been drawn with equal but with *unequal selection probabilities*. As a consequence, the sample will not be representative. To be able to draw valid conclusions, the response has to be corrected for this lack of representativity.

- The sample has not been drawn with probability sampling, but with self-selection. Therefore, selection probabilities are unknown, and unbiased estimates cannot be computed.

- The poll suffers from undercoverage. Therefore, parts of the target population are excluded. This may cause estimates to be biased.

- Some answers may be missing, or incorrect answers may have been replaced by a code for missing. To correct this, *holes* in the data set may have been filled with synthetic values. This is called *imputation*. If persons with missing answers differ from persons who answered the questions, imputation will lead to wrong conclusions. For example, if imputation of the mean (*holes* are filled with the mean of available values), the computed margins of error are too small, inadvertently creating the impression of very precise estimates.

- The data may be affected by *measurement errors*, that is, recorded answers differ from the correct answers. There is no guarantee that respondents have given the right answers to the questions. This can be caused by the respondents themselves. If a sensitive question is asked, they may give a socially desirable answer. If questions about the past are asked, respondents may have forgotten events. And if respondents do not want to think about the answer, they may just say "I don't know." Problems may also have been caused by the possibly bad ways in which questions were formulated. An example is omitting an answer option in a closed question.

- There will be *nonresponse* in the poll. If respondents differ from non-respondents (and this often happens to be the case), the researcher runs a serious risk of drawing wrong conclusions from the poll. Some kind of adjustment weighting has to be carried out to correct for nonresponse bias.

If the researcher discovers special patterns in the data set, he will attempt to come with an explanation. This may require more in-depth statistical analysis, such as regression analysis and factor analysis. Discussion of these techniques is outside of the scope of this book.

The remaining part of this chapter is devoted to exploratory analysis. Different groups of techniques are distinguished: from techniques for the analysis of the distribution of a single variable to techniques for the analysis of the relationship between two variables. Furthermore, one should realize that techniques for the analysis of quantitative variables cannot be used for the analysis of qualitative variables. Table 10.1 gives an overview of the techniques that are described. This overview is by

TABLE 10.1  Techniques for Explanatory Analysis

| Variables | Analysis of the Distribution | Analysis of the Relationship |
|---|---|---|
| Quantitative | • One-dimensional scatterplot<br>• Boxplot<br>• Histogram<br>• Summary table | • Two-dimensional scatterplot<br>• Correlation coefficient |
| Qualitative | • Bar chart<br>• Pie chart<br>• Frequency table | • Grouped bar chart<br>• Stacked bar chart<br>• Cross-tabulation |
| Mixed | | • Analysis of the distribution of a quantitative variable for each category of a qualitative variable |

no means complete. There are many more techniques available. This chapter is only about the most important ones.

There are graphical and numerical techniques for explanatory analysis. It is always a good idea to start with graphical techniques. The well-known proverb "one picture is worth a thousand words" certainly applies here. Graphs can summarize large amounts of data in a clear and well-organized way, thus providing an insight into patterns and relationships. If clear and simple patterns are found in the data, they can be summarized in a numerical overview.

Many of the techniques discusses here have been implemented in statistical software like *SPSS*, *SAS*, and *Stata*. These packages are generally large and expensive, as they offer much more analysis possibilities than just explanatory analysis. In this chapter, a simple and cheap approach was taken: The data were stored in the spreadsheet program *Excel*, and the free open source package *R* was used for analysis.

One small data set was used to illustrate all techniques in Table 10.1. This data set contains data about the working population of the small country of Samplonia. This country consists of two provinces: Agria and Induston. Agria is divided into three districts: Wheaton, Greenham, and Newbay. Induston is divided into four districts: Oakdale, Smokeley, Crowdon, and Mudwater. Table 10.2 contains a description of the contents of the data set.

To be able to analyze the data, they have to be entered in a computer. The spreadsheet program *Excel* was used here. Figure 10.1 shows part of the spreadsheet with data about the working population of Samplonia. Note that the first line contains the names of the variables.

TABLE 10.2   The Data Set for the Working Population of Samplonia

| Variable | Type | Values |
|---|---|---|
| District | Qualitative | Wheaton, Greenham, Newbay, Oakdale, Smokeley, Crowdon, and Mudwater |
| Province | Qualitative | Agria, Induston |
| Gender | Qualitative | Male, female |
| Age | Quantitative | From 20 to 64 |
| Employed | Indicator | Yes, no |
| Income | Quantitative | From 101 to 4497 |
| Ageclass | Qualitative | Young, middle, and old |

| | A | B | C | D | E | F | G |
|---|---|---|---|---|---|---|---|
| 1 | District | Province | Gender | Age | Employed | Income | Ageclass |
| 2 | Newbay | Agria | Female | 33 | Yes | 158 | Young |
| 3 | Wheaton | Agria | Male | 32 | Yes | 525 | Young |
| 4 | Mudwater | Induston | Female | 63 | Yes | 1442 | Old |
| 5 | Crowdon | Induston | Male | 29 | Yes | 1126 | Young |
| 6 | Wheaton | Agria | Female | 33 | Yes | 195 | Young |
| 7 | Mudwater | Induston | Male | 39 | Yes | 1559 | Middle |
| 8 | Oakdale | Induston | Male | 54 | Yes | 3760 | Old |
| 9 | Newbay | Agria | Male | 40 | Yes | 630 | Middle |
| 10 | Mudwater | Induston | Male | 53 | Yes | 2103 | Old |

FIGURE 10.1   The spreadsheet with the first 10 cases of the data set.

To prepare these data for analysis with R, they have to be saved as a CSV-file. The semicolon (;) is used as a field separator. The first line of the file should contain the names of the variables.

The free package R can be downloaded from the website www.r-project.org. After installing the package on your computer, and starting it, the CSV-file can be read with the function read.csv().

## 10.2  ANALYSIS OF THE DISTRIBUTION OF A QUANTITATIVE VARIABLE

The values of a quantitative variable are measured by means of a numerical question. First, three graphical techniques are described: the *one-dimensional scatterplot*, the *boxplot*, and the *histogram*; next, a numerical technique is described: the *summary table*.

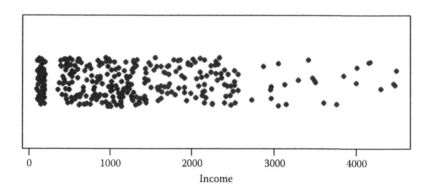

FIGURE 10.2   A scatterplot for one variable.

The *one-dimensional scatterplot* shows the distribution of a quantitative variable in its purest form. A horizontal axis has a scale that reflects the possible values of the variable. The individual cases are plot as points alongside this axis. Figure 10.2 contains an example of a one-dimensional scatterplot. It shows the distribution of the variable income for the working population of Samplonia.

Note that the points have different vertical distances to the X-axis. These distances are completely random. It is called *jitter*. The jitter has been applied on purpose to prevent points from overlapping each other. So all points remain visible.

There are different aspects of the one-dimensional scatterplot worth paying attention to. It is difficult to give general guidelines, because each variable is different, and there can always be unexpected patterns. Still, here are some aspects that may be of interest:

- *Outliers*: Are there values that are completely different from the rest? Such values manifest themselves as isolated points in the graph. Careful attention should be paid to these points. Maybe an error was made when processing the answers to the question. It is, however, also possible that the value was correct, and this was just a very special person in the sample.

- *Grouping*: Are the values more or less equally spread over a certain interval? Or can various groups of values be distinguished? If there are several separate groups, this may be an indication that the population consists of different subpopulations, each with their own behavior. Then, it may be better to analyze these groups separately.

- *Concentration*: Is there in an area with a high concentration of values? Maybe all values concentrate around a certain point. Then, it may be interesting to characterize this point. There are other analysis techniques to do this.

There are no outliers in Figure 10.2. There seems to be a separate group with very low incomes. Further analysis should show what kind of group this is. The incomes do not seem to concentrate around a central value. The distribution is skewed to the right with many low incomes and a few high incomes.

A second graphical technique for displaying the distribution of a quantitative variable is the *boxplot*. It is sometimes also called the *box-and-whisker plot*. The boxplot is a schematic presentation of the distribution of a variable. A box represents the middle half of the distribution. The *whiskers* represent the tails of the distribution. Figure 10.3 contains the boxplot for income of the working population of Samplonia.

The box denotes the area containing the middle half (50%) of the values. The vertical line in the box is the *median* of the distribution. Lines (*whiskers*) extend from the left and right of the box. These lines run to the values that are just within a distance of 1.5 times the length of the box. All values further away are drawn as separate points. These points are *outliers*.

A boxplot is useful for exploring the symmetry of a distribution. For a symmetric distribution, the median is in the middle of the box, and the whiskers have the same length. The boxplot is particularly powerful for detecting outliers. But be careful. For example, the outliers in Figure 10.3 are not really outliers. They only seem outliers because the distribution is skewed to the right. So there is a long tail on the right.

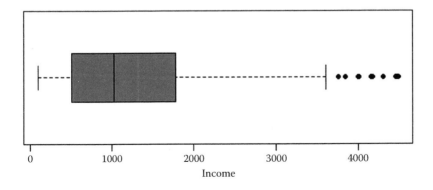

FIGURE 10.3   A boxplot.

The *histogram* also uses a horizontal line representing the range of possible values of the variable. The line is divided into a number of intervals of equal length. Next, the number of values in each interval is counted. A rectangle is drawn above each interval. The widths of the intervals are the same, but the heights are proportional to the numbers of observations. The rectangles are drawn adjacent to each other. Figure 10.4 contains a histogram for the incomes of the working people in Samplonia.

A point for consideration is the number of intervals in which the range of possible values is divided. If there are just a few intervals, one only gets a very coarse idea of the distribution. Much details are hidden. If there are many intervals, there will be so many details that the global picture is hidden. An often used rule of thumb is to take the number of intervals approximately equal to the square root of the number of values, with a minimum of 5 and a maximum of 20.

A histogram can be used to determine whether a distribution is symmetric with a peak in the middle. If this is the case, the distribution more or less resembles the *normal distribution*. Consequently, it can characterize the distribution by a few numbers (mean and standard deviation).

Figure 10.4 contains an example of a histogram. It shows the distribution of the incomes of the working people in Samplonia. The distribution is certainly not symmetric. There are many people with a small income and

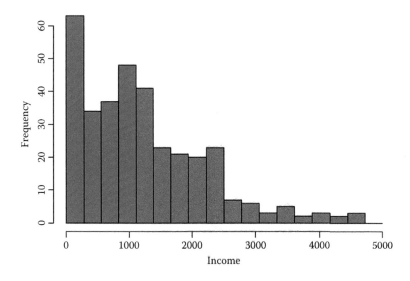

FIGURE 10.4    A histogram.

only a few people with a large income. One encounters a pattern like this often when displaying the distribution of quantities or values.

The distribution in Figure 10.4 also seems to have more than one peak. This can happen when several groups are mixed each having its own distribution. It could be a good idea to attempt to identify these groups and study each of them separately. It will be clear that for a mix of distributions, it will be difficult to characterize data in two numbers (means and standard deviation).

If the distribution of the values of the quantitative variables is more or less normal (symmetric, bell shaped), it can be summarized in the so-called summary table. Such a summary could contain the following quantities:

- *Minimum*: This is the smallest value in the data set.

- *Maximum*: This is the largest value in the data set.

- *Mean*: This is the central location of the distribution. All values are concentrated around this location.

- *Standard deviation*: This is a measure of the spread of the values. The more the values vary, the larger the standard deviation will be. The standard deviation is 0 if all values are equal.

- *Rule-of-thumb interval*: This is the interval containing approximately 95% of the values, provided the distribution is more or less normal (a bell-shaped distribution). The lower bound of the interval is obtained by subtracting two times the standard deviation from the average. The upper bound of the interval is obtained by adding two times the standard deviation to the average.

The distribution of the incomes of the working population of Samplonia is skewed and has several peaks. It is certainly not normal. Therefore, it is not possible to summarize it in the summary table. If only the data of the 58 working males in the province of Agria was taken into account, the distribution would be much more normal. The summary table of this distribution is presented in Table 10.3.

The incomes of the working males in Agria vary between 353 and 841. They concentrate around a value of 551.2. A standard deviation of 119.3 leads to a rule-of-thumb interval running from 312.6 to 789.8. So, 95% of the incomes are between 312.6 and 789.8.

TABLE 10.3 Summary Table

| Variable | Income |
| --- | --- |
| Number of cases | 58 |
| Minimum | 353 |
| Maximum | 841 |
| Average | 551.2 |
| Standard deviation | 119.3 |
| Rule-of-thumb interval | (312.6; 789.8) |

## 10.3 ANALYSIS OF THE DISTRIBUTION OF A QUALITATIVE VARIABLE

Qualitative variables are measured with a closed question. Only a few techniques are available for the analysis of the distribution of a qualitative variable. The reason is that no computations can be carried out with the values of qualitative variables. Their values are just labels, or code numbers for labels. These labels denote categories. They divide the population into groups. The only thing one can do is counting the number of people in a group.

Two graphical techniques are described here: the *bar chart* and the *pie chart*. One numerical technique will be presented: the *frequency table*.

The first graphical technique is the *bar chart*. It presents the groups as bars, in which the lengths of the bars reflect the numbers of persons in the groups. To avoid confusion with a histogram, it is better to draw the bars of a bar chart horizontally and to have some space between the bars.

If the categories of a qualitative variable have no natural order (it is a *nominal variable*), the bars can be drawn in any order. One can take advantage of this by ordering the bars from small to large (or vice versa). This makes it easier to interpret the graph.

Figure 10.5 contains an example of a bar chart with ordered bars. It shows the numbers of employed in the seven districts of Samplonia. There is a bar for each district, and the length of the bar reflects the number of employed in the district. It is clear from the bar chart that two districts (Smokeley and Mudwater) have substantially more employed than the other districts. Also note that two districts (Oakdale and Newbay) have a small number of employed.

A second graphical technique for the analysis of the distribution of a qualitative variable is the *pie chart*. This type of graph is particularly

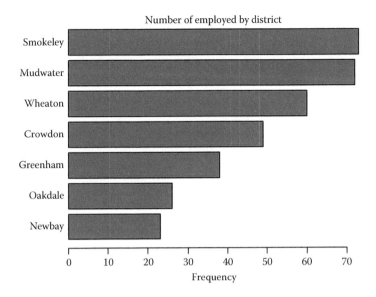

FIGURE 10.5   A bar chart.

popular in the media. The pie chart is a circle (pie) that is divided into as many sectors as the variable has categories. The area of the sector is taken proportional to the number of people in the corresponding category. Figure 10.6 contains an example. It shows the numbers of employed persons per district. So it contains the same information as Figure 10.5.

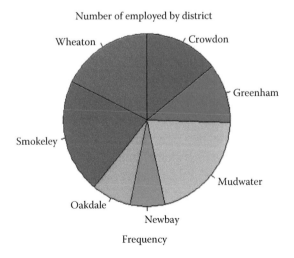

FIGURE 10.6   A pie chart.

TABLE 10.4   A Frequency Table

| Category | Frequency | Percentage (%) |
|----------|-----------|----------------|
| Wheaton | 60 | 17.6 |
| Greenham | 38 | 11.1 |
| Oakdale | 26 | 7.6 |
| Newbay | 23 | 6.7 |
| Smokeley | 73 | 21.4 |
| Crowdon | 49 | 14.4 |
| Mudwater | 72 | 21.1 |
| Total | 341 | 100.0 |

Pie charts are sometimes more difficult to interpret than bar charts. If there are many parts, which all have approximately the same size, it is hard to compare these parts. This is much easier in a bar chart. It may help one to order the sectors in ascending or descending size.

The numerical way the present the distribution of a qualitative variable is the *frequency table*. This is a table with for each category the number and the percentage of people that category. Table 10.4 contains the frequency distribution of the numbers of employed by district in Samplonia.

As already mentioned, it is not possible to do calculations with a qualitative variable. Therefore, an average value cannot be computed. If it is really important to characterize the distribution in one number, the mode could be determined. The *mode* is defined as the value that appears most often in a data set. For example, for the number of employed per district, the category Smokeley is the mode. There are 73 persons in this category (see Table 10.4), and this corresponds to 21.4% of all employed.

## 10.4 ANALYSIS OF THE RELATIONSHIP BETWEEN TWO QUANTITATIVE VARIABLES

To analyze the relationship between two quantitative variables, there is one very popular technique: the *two-dimensional scatterplot*. Every person in the sample is presented by a point in a coordinate system. The horizontal coordinate is equal to the value of one variable, and the vertical coordinate is equal to the value of the other variable. Thus, a cloud of points is created.

If the cloud does not show a clear structure (it is a random snow storm), there is no relationship between the two variables. If there is some clear pattern, there is a relationship between the two variables. Then it is a good idea to further explore the nature of this relationship.

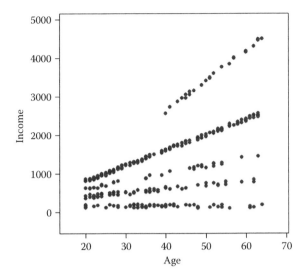

FIGURE 10.7    A scatterplot for two variables.

The most extreme form of relationship is the one in which all points are on a straight line. In this case, the value of one variable can be predicted exactly from the value of the other variable. Also other aspects of a two-dimensional scatterplot could be interesting. For example, it will not be difficult to detect outliers, or a group of points that behave differently from the rest.

Figure 10.7 shows a two-dimensional scatterplot of the variables age and income in the working population of Samplonia. The points have a clear structure. First, points are divided in several different groups. For example, there is a smaller group of points representing people of age 40 and older and who have very high incomes. Income increases with age in this group. There is also a group with very low incomes, where income does not increase with age.

If the points fall apart in separate groups, it could be interesting to characterize these groups. It may help us to include more variables in the analysis. If there is another quantitative variable, its value can be indicated by the size of the points (by taking the size of the point proportional to the value of this variable). If the third variable is a qualitative variable, points can be given colors that correspond to the categories of this variable.

The structure of the points can be so complex that it is not possible to summarize them in a few numbers. However, one may always hope to encounter a situation in which such a summary is possible. One situation is that in which all point are more or less on a straight line. It means there is a linear relationship. Such a relationship can be summarized by a *correlation coefficient* and a *regression line*.

The *correlation coefficient* attempts to express the strength of the relationship in a value between −1 and +1. Note that the correlation coefficient only works well for linear relationships. For example, if all points form a parabolic curve, the relationship is still very strong, but this will not show in the value of the correlation coefficient.

The value of the correlation coefficient can vary between −1 and +1. The correlation coefficient is 0 if there is no relationship at all between the two variables. If there is a perfect linear relationship between the two variables, the correlation coefficient is either −1 (for a downward trend) or +1 (for an upward trend). A correlation of −1 or +1 means that the value of one variable can be predicted exactly from the value of the other variable.

It is not meaningful to compute the correlation coefficient before having looked at the two-dimensional scatterplot. Figure 10.7 illustrates this. The correlation coefficient for all points turns out to be 0.568. This value is not very high. So there only seems to be a weak relationship. However, relationships within the various groups are strong. For example, the correlation coefficient within the high income group is 0.963, which indicates an almost perfect linear relationship in this group.

If there is a strong relationship between two variables, and all points are approximately on a straight line, this relationship can be summarized by means of a *regression line*. The computations for such a regression line are outside the scope of this publication. Fortunately, most statistical software packages can do it. It is not meaningful to compute the regression line for the scatterplot in Figure 10.7, because the points do not form a straight line. The situation is different in Figure 10.8. This is the two-dimensional scatterplot of the variables age and income for the working males in the province of Agria. There is clearly a linear relationship. The correlation coefficient is a good indication for the strength of the relationship. Its value turns out to be 0.963, which is almost equal to +1. So there is a strong relationship. The expression for the regression line is

$$205.493 + 9.811 \times \text{Age}$$

This means someone's income can be predicted by multiplying his age by 9.811 and adding 205.493 to the result. For example, if a male in Agria has an age of 50 years, his income should be something like $205.493 + 9.811 \times 50 = 696.043$.

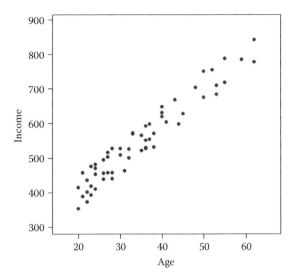

FIGURE 10.8    A two-dimensional scatterplot for two variables with a linear relationship.

## 10.5 ANALYSIS OF THE RELATIONSHIP BETWEEN TWO QUALITATIVE VARIABLES

The possibilities for analysis of the relationship between qualitative variables are limited, as it is not possible to do meaningful calculations with these types of variables. For graphical analysis, extensions of the bar chart can be used: the *grouped bar chart* and the *stacked bar chart*. Furthermore, a set of *pie charts* can be produced. And there is the quantity *Cramérs V* to numerically measure the strength of the relationship between two qualitative variables.

A *grouped bar chart* consists of a set of bar charts. A bar chart is drawn of one variable for each category of another variable. All these bar charts are combined into one graph. Figure 10.9 contains an example of a grouped bar chart. It is the age distribution (in age categories) by district for the working population of Samplonia. The bars have been drawn horizontally. Again, this is done to avoid confusion with histograms.

The grouped bar chart contains a bar chart for the age distribution in each district. For example, one can see that there are no young people in Oakdale and no old people in Newbay. Of course, the roles of the variables can be switched, resulting in a bar chart of the distribution over the districts for each age category. This helps us to detect other aspects of the relationship between age and district.

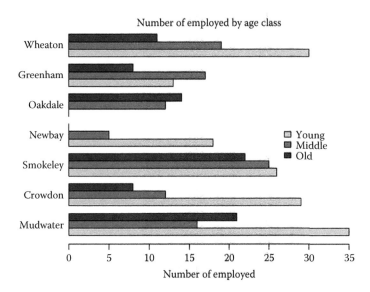

FIGURE 10.9   A grouped bar chart.

Is the shape of the bar chart for the one variable different for each category of the other variable, then it may be concluded that there is some kind of relationship between the two variables. To further investigate the nature of this relationship, a closer look can be taken at each bar chart separately.

Note that the chart will not reveal every single aspect of the relationship between the two variables. It will show only certain aspects. For example, it is difficult to see in Figure 10.9 which district has a maximum number of employed persons. For this, one would have to combine all the bars for each district into a single bar. It is also difficult to see whether an age category is under- or overrepresented in a specific district. It is hard to say whether the percentage of young people in Smokeley is larger or smaller than the percentage of young people in Mudwater. It is easier to compare numbers of people in the various age categories. For example, one can conclude that Smokely has the largest group of elderly.

Another way to show bar charts of one variable for each category of another variable is the *stacked bar chart*. The bars of each bar chart are not drawn below each other (like in Figure 10.9) but stacked onto each other. Figure 10.10 shows an example of a stacked bar chart. The same data are used as in Figure 10.9.

The total length of all bars is proportional to the number of employed in the various districts. Each bar is divided into segments the lengths of which reflect the distribution of the age categories.

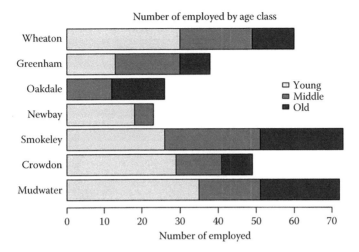

FIGURE 10.10   A stacked bar chart.

What is there to see in a stacked bar chart? It is clear which category of the one variable (district) is the smallest (Newbay), and which one is the largest (Smokeley). Furthermore, one can get a good idea which category of the other variable is under- or overrepresented within a category of the one variable. For example, it is clear that there are no young employed in Oakdale, and no old employed in Newbay. Comparing the age distributions of two districts is difficult.

Comparing distributions for the other variable within categories of the one variable is easier if the counts in the bars are replaced by percentages. This will produce a stacked bar chart in which each bar has the same length (100%) and the various segments in the bars represent the distribution of the other variable in percentages. See the example in Figure 10.11.

A chart like this one makes it possible to compare relative distributions. For example, one cannot only compare the percentages of young employed in the various districts, but also the percentages old employed. It is clear that relatively more old employed people live in Smokeley and Mudwater than in Wheaton or Greenham. Relatively many young employed people live in Newbay and Smokeley, whereas middle-aged employed people seem to be well represented in Greenham and Oakdale.

It was already mentioned that the pie chart can be an alternative for the bar chart. Also for the analysis of the relationship between two qualitative variables pie charts can be used. The bar charts in Figure 10.9 could simply be replaced by pie charts. Such a graph will become even more informative

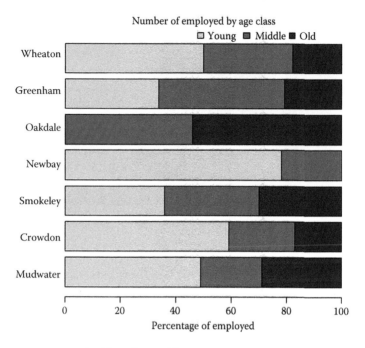

FIGURE 10.11    A stacked bar chart with percentages.

if the sizes of the pie charts are taken proportional to the number of people on which they are based (the number of people in the category of the other variable). Figure 10.12 shows an example.

The areas of the circles are proportional to the numbers of employed in the corresponding districts. One can see which district is the largest (Mudwater) and which district is the smallest (Newbay) in terms of number of employed people. One can also see the age distribution within each district. It is hard,

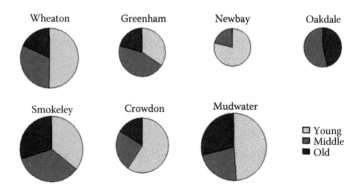

FIGURE 10.12    Pie charts.

however, to compare the age distributions. It is also not so easy to compare the sizes of the same age group in different districts.

So there are various instruments to explore the relationship between two qualitative variables in a graphical way. There is no single technique that performs best. Each technique has strong points and weak points. The best thing to do is to try all techniques.

A numerical overview of the combined distribution of two qualitative variables can be obtained by making a *cross-tabulation*. Table 10.5 is an example. It shows the cross-tabulation of the variables district and age class of the working population in Samplonia.

Interpretation of counts in a small table is doable, but it becomes harder as the number of rows and columns increases. What may help is to replace counts by percentages. Percentages can be computed in various ways: percentages of the table total, row percentages, and column percentages. As an example, Table 10.6 contains row percentages. This means all

TABLE 10.5   A Cross-Tabulation

| District | Young | Middle-Aged | Old | Total |
|---|---|---|---|---|
| Wheaton | 30 | 19 | 11 | 60 |
| Greenham | 13 | 17 | 8 | 38 |
| Oakdale | 0 | 12 | 14 | 26 |
| Newbay | 18 | 5 | 0 | 23 |
| Smokeley | 26 | 25 | 22 | 73 |
| Crowdon | 29 | 12 | 8 | 49 |
| Mudwater | 35 | 16 | 21 | 72 |
| Total | 151 | 106 | 84 | 341 |

TABLE 10.6   A Cross-Tabulation with Row Percentages

| District | Young (%) | Middle-Aged (%) | Old (%) | Total (%) |
|---|---|---|---|---|
| Wheaton | 50.0 | 31.7 | 18.3 | 100.0 |
| Greenham | 34.2 | 44.7 | 21.1 | 100.0 |
| Oakdale | 0.0 | 46.2 | 53.8 | 100.0 |
| Newbay | 78.3 | 21.7 | 0.0 | 100.0 |
| Smokeley | 35.6 | 34.2 | 30.1 | 100.0 |
| Crowdon | 59.2 | 24.5 | 16.3 | 100.0 |
| Mudwater | 48.6 | 22.2 | 29.2 | 100.0 |
| Total | 44.3 | 31.1 | 24.6 | 100.0 |

percentages in a row add up to 100%. In this way, the age distribution is obtained within each district.

This table makes clear that young employed people are overrepresented in Newbay (78.3%), and that relatively many old employed people live in Oakdale (53.8%).

If there is no relationship between the row and column variable, the relative distribution of the column variable will be more or less the same in each row. And vice versa, the relative distribution of the row variable will be more or less the same in each column. The larger the differences between the relative distributions, the stronger the relationship is.

There are numerical quantities that attempt to express the strength of the relationship between two qualitative variables in one number. One way to do this is to compute *Cramérs V*. This quantity is based on the well-known *chi-square statistic*. If this test statistic is close to zero, there is (almost) no relationship. The larger its value, the stronger the relationship. But what is large? A problem of the chi-square statistic is that its value does not only depend on the strength of the relationship, but also on the numbers of rows and columns, and on the number of observations. So it is difficult to interpret its value. Therefore, other quantities have been defined that are independent of these numbers. One such quantity is *Cramérs V*. The value of *Cramérs V* is always between 0 and 1. A value of 0 means a total lack of relationship. A value of 1 means a perfect relationship. A rule of thumb is to consider values below 0.3 as a weak relationship, values between 0.3 and 0.7 as a moderate relationship, and values of more than 0.7 as a strong relationship.

The value of *Cramérs V* is 0.268 for the data in Table 10.5. So one can conclude there is a weak relationship between district and age distribution in the working population of Samplonia.

## 10.6 ANALYSIS OF THE RELATIONSHIP BETWEEN A QUANTITATIVE AND A QUALITATIVE VARIABLE

There are no specific techniques for the analysis of the relationship between a quantitative and a qualitative variable. What one can do is take a technique for the analysis of the distribution of a quantitative variable and apply it for every category of the qualitative variable. Two graphical techniques will be described, one based on the one-dimensional scatterplot, and the other on the boxplot. It is also possible to make a numerical overview based on the summary table.

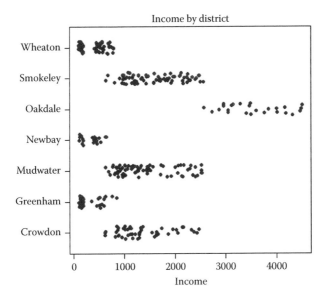

FIGURE 10.13   Scatterplots for one variable.

The *one-dimensional scatterplot* can be used to study a possible relationship between a qualitative variable and a quantitative variable. The idea is to make a one-dimensional scatterplot of the quantitative variable for each category of the qualitative variable and to combine all these scatterplots in one graph. To avoid points from overlapping each other, vertical jitter is added in each scatterplot.

Figure 10.13 gives an example. The scatterplot shows the relationship between the variables income and district for the working population of Samplonia.

It is clear that the incomes are low in Wheaton, Newbay, and Greenham. In each of these three districts, there seems to be two groups: one group with very low incomes and the other with somewhat higher incomes. The incomes are clearly the highest in Oakdale. There is only one group. The other three districts (Smokeley, Mudwater, and Crowdon) take a middle position with respect to income.

The second graphical technique for the analysis of the relationship between a quantitative variable and a qualitative variable is a set of boxplots: draw a boxplot of the quantitative variable for each category of the qualitative variable. To be able to compare the boxplots, the same scale must be used for all boxplots.

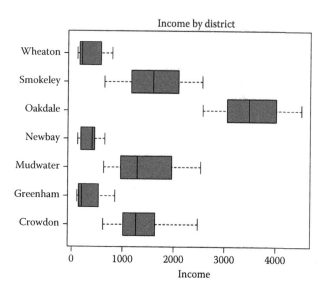

FIGURE 10.14   Boxplots.

Figure 10.14 contains the boxplots for the analysis of the relationship between income and district for the working population of Samplonia. The graph shows clear differences between the income distributions. There are districts with low incomes (Wheaton, Newbay, and Greenham), and there is also a district with very high incomes (Oakdale). The income distributions of the other three districts (Smokeley, Mudwater, and Crowdon) are somewhere in between. The income distributions of these three districts are more or less the same.

A similar approach can be used for a numerical analysis of the relationship between a quantitative variable and a qualitative variable: do a numerical analysis of the quantitative variable for each category of the qualitative variable.

Table 10.7 presents an example. It is a table that contains for each district a number of characteristics of income: number of cases, minimum value, maximum value, average value, and standard deviation. Other quantities, like the rule-of-thumb interval and the median, could also be included.

Table 10.7 shows that the income distribution in Oakdale is very different from the income distributions in the other districts. The standard deviation of income is in Wheaton, Greenham, and Newbay lower

TABLE 10.7    A Numerical Overview of the Variable Income

| District | Cases | Minimum | Maximum | Average | Standard Deviation |
|----------|-------|---------|---------|---------|--------------------|
| Wheaton | 60 | 101 | 787 | 356 | 234 |
| Greenham | 38 | 102 | 841 | 324 | 219 |
| Oakdale | 26 | 2564 | 4497 | 3534 | 586 |
| Newbay | 23 | 115 | 648 | 344 | 167 |
| Smokeley | 73 | 635 | 2563 | 1607 | 518 |
| Crowdon | 49 | 612 | 2471 | 1356 | 505 |
| Mudwater | 72 | 625 | 2524 | 1440 | 569 |
| Total | 341 | 101 | 4497 | 1234 | 964 |

than in the other districts. Apparently, incomes do not differ so much in these three districts.

Again, it is advised to first have a graphical look at the distributions of the quantitative variable before summarizing them in a numerical overview. The numerical summary is only meaningful if the distributions of the quantitative variable are more or less normal (bell-shaped). For example, some income distributions within districts are rather skew. If one would compute a rule-of-thumb interval for income in Wheaton, the lower bound is negative. This is silly, as all incomes are positive. So the rule-of-thumb interval does not work here.

## 10.7 SUMMARY

The analysis of the collected data should reveal interesting and relevant patterns in the data. Before starting the analysis, the researcher should always check the data for errors. Graphical and numerical techniques can be used for this, of course, detected errors should be corrected.

After the data have been cleaned, the analysis can begin. The first step should always be an exploratory analysis. Various graphical and numerical tools can help the research to discover new, unexpected patterns and relationships in the data.

Exploratory analysis aims at investigating the collected data, and nothing more. The conclusions only relate to the data set. No attempt is made to generalize from the response of the poll to the target population.

The exploratory analysis can be followed by an inductive analysis. This can take the form of computing estimates of population characteristics, and of testing hypotheses about the target population. Note that the conclusions of an inductive analysis are based on sample data and, consequently, have an element of uncertainty. Therefore, margins of error should be taken into account.

There are tools for numerical and graphical analysis of data. It is a good idea to start with a graphical analysis technique. They are more powerful for detecting unexpected patterns in the data. If clear patterns have been found, they can be summarized in numerical overviews.

CHAPTER **11**

# Publication

## 11.1 THE RESEARCH REPORT

The ultimate goal of a poll will be to get more insight in the population that is investigated. So it is obvious to make a research report. The purpose of this report is twofold. In the first place, it must contain the results of the analysis of the collected data and thus provide new insights in the target population. In the second place, the research report must contain an account of which data were collected, how they were collected, how the data were processed, how they were analyzed, and how the conclusions were reached.

The report has to satisfy two important conditions. In the first place, the report must be readable. Technical jargon should therefore be avoided. Also readers without a statistical background must be able to read and understand it. In the second place, the researcher is accountable for the way in which he conducted his poll. Other experts in the field of survey research must be able to check whether the researcher followed the methodological principles, and whether he or she drew scientifically sound conclusions about the target population.

The research report should be written in a concise and professional writing style. The report should be objective and neutral, and not enforce a specific opinion. The use of common language should be avoided. It should be written in the passive form, and not to use the *you-style*. Unfamiliar terms and symbols must be avoided.

If the outcomes of a poll are for a wide audience, it is especially important to present the results in such a way that everybody can understand them. Use of graphs can help in this respect. A graph can be a very powerful instrument for showing and conveying the *message* in the data. Design of graphs requires careful attention. Shortcomings in the design may easily mislead viewers of the graph. This chapter shows how to distinguish good graphs from bad graphs.

The remainder of this chapter consists of two parts: Section 11.2 describes the structure of the research report. It also explains which methodological information must be included. Section 11.3 is about the use of graphs in reports. It provides guidelines for making good graphs.

## 11.2 THE STRUCTURE OF THE RESEARCH REPORT

So a research report must be produced for each poll. How should this report look like? What should be in it? There are various ways in which structure can be given to a research report. The research report must at least have a subject-matter part and a methodological part. The subject-matter part describes the analysis of the poll data and conclusions that were drawn from the analysis. The methodological part should describe how the poll was set up and carried out. The structure proposed in this section is just one way to do this. One can think of other ways to write a report. This structure is, however, often used in practice.

An often suggested structure is to divide the report in the following sections:

- The executive summary
- The methodological account
- The outcomes of the analysis
- The conclusions
- Literature
- Appendices

These parts are described in more detail in the following subsections.

### 11.2.1 The Executive Summary

The *executive summary* is a short description of the poll in a manner that is readable and understandable for everyone. It consists of two parts: the *research question* and the *conclusions*.

The *research question* is a clear, focused, and concise question about a practical issue. It is the question around which the poll was centered.

It should describe what the practical issue is in subject-matter terms, and how the poll will provide the answer to the question raised.

It should be clear from the research question who commissioned the poll, and also who sponsored it. Possible interests of these organizations in specific outcomes of the poll should be mentioned.

After having described the research question, an overview must be given of the most important conclusions of the poll. Such conclusions must be supported by the collected data. The conclusions should contribute to answering the research question.

It is important that everyone (particularly commissioners and sponsors, but possibly also the media and the general public) understands the conclusions. It must be made clear how far the conclusions reach. What is the target population to which they refer? And what is the reference date of the survey? The executive summary should also indicate how accurate the conclusions are. What are the margins of error? Was there any nonresponse, and what was its effect on the outcomes?

The executive summary should be short and should consist of no more than a few pages. Particularly, the conclusions should be concise but placed in the right context. There should only be conclusions, and no arguments leading to the conclusions.

The executive summary must be readable and understandable by the commissioner of the poll. The commissioner must be able to implement the conclusions in concrete policy decisions. There is no place here for statistical jargon or mathematical expressions.

## 11.2.2 The Methodological Account

The methodological account is the methodological part of the report. This is an accurate description of the how the poll was designed and carried out. This account should provide sufficient information to determine whether the conclusions were drawn in a scientifically sound way. The conclusions should be supported by the collected data.

The methodological account must address a number of aspects of the poll. There are various organizations that have compiled lists of what should be included in this account. The following list was obtained by combining the lists of these organizations:

- The National Council on Public Polls (NCPP, www.ncpp.org) is an American association of polling organizations. Its mission is to set the highest professional standards for public opinion pollsters and to advance the understanding, among politicians, the media, and general public, of how polls are conducted and how to interpret poll results. The National Council on Public Polls has published *Principles for Disclosure*. These principles should provide an adequate basis for judging the reliability and validity of the poll results reported.

- The American Association of Public Opinion Research (AAPOR, www.aapor.org) is the leading association of public opinion professionals in America. The AAPOR community includes producers and users of survey data from a variety of disciplines. AAPOR has developed a Code of Ethics. This code sets standards for ethical conduct of polling organizations. It also contains recommendations on best practices for polls. The Code of Ethics contains *Standards for Disclosure*. They describe the information that should be provided in research reports.

- The World Association for Public Opinion Research (WAPOR, www. wapor.org) is the leading international organization promoting the use of science in the field of public opinion research. It has developed a code of ethics and professional standards. This code contains rules of practice for research reports.

- European Society for Opinion and Marketing Research (ESOMAR, www.esomarg.org) is an international organization for encouraging, advancing, and elevating market research and opinion research worldwide. Originally it was a European organization. Together with World Association for Public Opinion Research, it has developed guidelines that set out the responsibilities of researchers to conduct opinion polls in a professional and ethical way. These guidelines contain requirements for publishing poll results.

- The Dutch-speaking Platform for Survey Research (NPSO, www. npso.net) developed together with Statistics Netherlands (CBS, www.cbs.nl), and the Dutch-Flemish Association for Investigative Journalism (VVOJ, www.vvoj.nl) a checklist for polls. This checklist helps users of poll results to assess the quality of a poll. This checklist can only be used if a proper research report is available. The checklist is described in more detail in Chapter 12.

There is a large overlap between the lists of items that should be included in the poll documentation. Consequently, it is not so difficult to make a combined list. The following items should be included:

1. *Organizations involved*: It is important to know who commissioned or sponsored a poll, because such an organization may have an interest in certain outcomes. If data collection for the poll was carried out by an organization other than the sponsoring organization, this organization must also be mentioned.

2. *Target population*: The target population is the group of people from which the sample was drawn and to which the outcomes of the poll refer. There must be a clear definition of the target population. It must always be possible to decide in practical situations whether or not a person belongs to the target population.

3. *Questionnaire*: A good questionnaire is of vital importance. Indeed, practice has shown that it is easy to influence the outcomes of a poll by manipulating the texts of the questions and the order of the questions. It must be possible to determine the quality of the questionnaire. Therefore, the questionnaire must be available for inspection.

4. *Sample selection*: How was the sample selected? It should be probability sample with known selection probabilities. The preferred way of doing this is drawing a simple random sample without replacement (and with equal probabilities). Note that if no probability sampling was applied, selection probabilities are unknown, making it impossible to compute valid estimates. This is, for example, the case for a quota sampling, and for self-selection sampling. Many online polls are based on random samples from web panels. This only works if the web panel is representative for the target population. Therefore, it is important to describe how this panel was constructed. Preferably, it should be based on a random sample from the target population, and not on self-selection.

5. *Sample size*: This is the *gross sample size*. It is the sample size as intended by the researcher. It is the result of a sample size calculation. Not everyone in the gross sample will respond. The number of respondents is also called the *net sample size*.

6. *Dates*: A poll describes the status of the population at a certain point in time. Therefore, it should be made clear at which dates the data were collected.

7. *Mode of data collection*: The mode of data collection (face-to-face, telephone, mail, online) has an impact on the quality of the estimates

of the poll. Therefore, the mode of data collection should be described. It should also be made clear whether some form of computer-assisted interviewing (CAPI, CATI) was implemented, whether the routing through the questionnaire was enforced, and whether the questionnaire contained checks.

8. *Response rate*: The response rate must be reported. The response rate is obtained by dividing the net sample size by the gross sample size, and multiplying the result by 100. The response rate can be seen as a quality indicator: the lower the response rate, the higher the risk of substantially biased estimates.

9. *Weighting*: The representativity of a poll can be affected by under-coverage, self-selection, and nonresponse. To reduce the effects of the lack of representativity, weighting adjustment must be applied. Not every weighting procedure is effective. It depends on the weight variables used. Therefore, the weighting procedure should be described.

10. *Margins of errors*: Even the outcomes of a perfect poll (random sampling, no nonresponse) are only estimates of population characteristics. It is important to stress this by specifying the margins of errors. This makes it possible for users of the poll data to distinguish real patterns from sampling artifacts.

### 11.2.3 The Outcomes of the Analysis

The third part of the research report describes the analysis of the collected data. It should start with an exploratory analysis of the data. This exploratory analysis must provide insight in each variable (target variable or auxiliary variable) separately.

The distribution of a variable can be analyzed with a graphical or with a numerical technique. Graphs are often easier to interpret and therefore may provide more insight (*one picture is worth a thousand words*). So the first choice could be for graphs. Tables with frequency distributions and other numerical details could possibly be included in an appendix.

The exploratory analysis may be followed by a more in-depth analysis in which the researcher attempts to describe and interpret the distribution of the target variables. He may also want to explore possible relationships between target variables, or between target variables and auxiliary variables. Again, one can choose between graphical and numerical techniques. The graphs give a global picture, whereas the numerical techniques provide the numerical details.

The researcher should not forget to take into account that all quantities have margins of error due to sampling. Were possible he must account for this uncertainty.

It may be better not to include too many technical details in the text, as this may affect its readability. The researcher could consider putting technical details in an appendix.

### 11.2.4 The Conclusions

The fourth part of the research report is devoted to the conclusions drawn from the poll. It is a translation back from the estimates to the practical issues that were investigated. This part should be more than just a listing of the conclusions, as in the first part. There should be more room for interpretation here and therefore somewhat more subjective conclusions. Nevertheless they may never contradict poll findings and should be based on the poll findings.

The conclusion could also be a hypothesis about the target population. Such a hypothesis should be tested in a new poll.

### 11.2.5 Literature

The fifth part of the research report should contain an overview of all literature used by the researcher. There could be two lists of publications:

- *Subject-matter literature*: These are the publications that were consulted about the subject-matter problem investigated in the poll.

- *Methodological literature*: These are the methodological publications that were consulted in order to design and conduct the poll in a scientific sound way.

### 11.2.6 Appendices

The appendices are for a number of things that are too big or too complex to be included in the regular text, and that are not really necessary to understand what was going on in the poll. It is information that is required to get a good assessment of the quality of the poll.

Here are a number of things that could be included in an appendix:

- The questionnaire

- An explorative analysis with, for each variable, a numerical summary of the distribution of the answers

- Expressions of estimation procedures, including expressions for computing correction weights

- Large tables with outcomes

- Letters (or e-mails) that have been sent to respondents, including reminders

## 11.3 USE OF GRAPHS

Poll results can be communicated in various ways. One obvious way to do it is in plain text. If there is a lot of information in the data, or if the information is complex, readers may easily lose their way. More structured ways of presenting the information are tables and graphs. Particularly for statistically less educated users, graphs have a number of advantages over text and tables. These advantages were already summarized by Schmid (1954):

- Graphs can provide a comprehensive picture. They make it possible to obtain a more complete and better balanced understanding of the problem.

- Graphs can bring out facts and relationships that otherwise would remain hidden.

- Use of graphs saves time as the essential meaning of a large amount of data can be visualized at a glance.

- Relationships between variables as portrayed by graphs are more clearly grasped and more easily remembered.

- Well-designed graphs are more appealing and therefore more effective for creating the interest of the reader.

Graphs can be used in polls in two ways. One way is to use them as tools for data analysis. Particularly, graphs can be very effective in an exploratory data analysis to explore data sets, to obtain insight, and to detect unexpected patterns and structures. Exploratory analysis of poll data is meant for the researchers themselves. Therefore, layout is less important. This type of use of graphs was already described in Chapter 10.

Another use of graphs is to include them in research reports. These graphs are meant for the reader of these reports. Most importantly, these graphs should be able to convey a message to statistically inexperienced readers. Therefore, the type of graph should be carefully selected. The visual display of the graph should be such that it reveals the message and not obscures it.

A graph can be a powerful tool to convey a message contained in poll data, particularly for those without experience in statistics. Graphs can be more meaningful and more attractive than tables with numbers. Not surprisingly, graphs are often encountered in research reports, and in the popular media (websites, newspapers, and television). Use of graphs is, however, not without problems. Poorly designed graphs may convey the wrong message. There are ample examples of such graphs. Problems are often caused by designers of graphs who lack sufficient statistical expertise. Unfortunately, they pay more attention to attractiveness of the graphic design than to its statistical content.

This section describes some popular types of graphs and show what the important design aspects are. Also some guidelines are presented that help to design graphs, and may help the readers of research reports to take a critical look at its graphs, and hence avoid drawing the wrong conclusions from the graphs.

### 11.3.1 Pie Chart

The pie chart is possibly the most popular type of graph. Pie charts are used for showing the distribution of answers to a closed question. There is a slice for each possible answer. The size of a slice reflects the numbers of respondents who selected this answer. So a large slice means many respondents in this category.

Figure 11.1 shows an example of a pie chart. People were asked for the party of their choice. The question was asked in a political poll just before the election in Scotland in 2016. The poll was conducted by the market research agency Survation. It is clear from this pie chart that the Scottish National Party (SNP) is by far the largest party.

The disadvantage of a pie chart is that it is not always easy to compare the sizes of the slices. For example, without further information it is hard to determine which party is larger: Labour (red) or the Conservatives (blue).

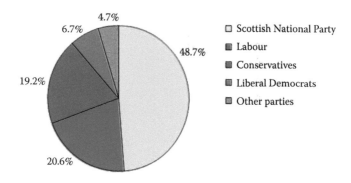

FIGURE 11.1    A pie chart for a Scottish poll. (From Survation, May 2016.)

To assist readers with their comparison, percentages have been included in the chart.

It would be easier for the viewer of the graph to have the labels of the categories in or close to the slices. Unfortunately, there is often not enough room to do this. Therefore, a legend is included, which links colors to labels. From a cognitive point of view, this makes interpreting the chart more cumbersome.

To be able to distinguish the slices, they need to have different colors. This is not so difficult in the case of the Scottish poll. In fact, for each party its own color was used. This becomes more complicated if there are many slices. How to see to it that all colors are different, and moreover, that they have the same intensity?

Figure 11.2 contains a pie chart for a political poll in The Netherlands. The poll is called Peilingwijzer. It shows the estimates for the Dutch political parties in April 2016. As is clear from the graph, there are many parties in The Netherlands. This makes it difficult to make a readable pie chart.

There are several parties with approximately the same size (e.g., SP, CDA, and D66). Without the percentages, it is hard to compare them. An attempt was made to use the own colors of the parties. As some of these colors are more or less the same, it requires some effort to link parties to slices.

Pie charts are not the best way to display poll results for a closed question. If a researcher insists on using pie charts, he should restrict it to questions with no more than six possible answers. So, not more than six slices should be considered. And focus should be on comparing a single slice with the total (and not with another slice).

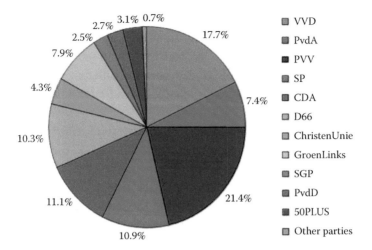

FIGURE 11.2   A Dutch poll. (From Peilingwijzer, April 2016.)

## 11.3.2  Bar Chart

A bar chart is a good alternative for a pie chart. Bar charts do not have the disadvantages of pie charts. Pie charts can be more attractive in terms of shape and colors, but the less spectacular bar charts are more effective. So bar charts are better for conveying the message in the data.

A bar chart contains bars. Each bar corresponds to a possible answer. The length of a bar reflects the number of respondents in the corresponding category. Figure 11.3 shows an example of a bar chart. The same data were used as in the pie chart in Figure 11.1. It is the question in a poll about voting intention at the Scottish election of May 2016.

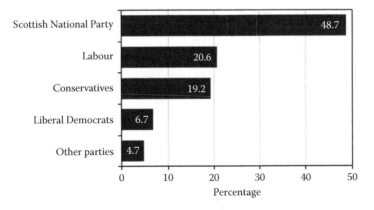

FIGURE 11.3   A bar chart for a Scottish poll. (From Survation, May 2016.)

The bar chart has horizontal bars, and the bars have some spacing between them. Both properties help us to distinguish a bar chart from a histogram (see Section 10.1). By drawing the graph in this way, it is no longer necessary to use different colors for different bars. One color suffices. This arrangement provides ample space for each label close to its corresponding bar.

The designer of this chart must not forget to provide the horizontal axis with proper scale values. There should not be too few and too many values. And it must be clear what the meaning of the values is. So the unit of measurement must be mentioned. The bar chart in Figure 11.3 also contains some gridlines that help us to determine the lengths of the bars. These gridlines should have not a too prominent color.

Note that the bars in Figure 11.3 are ordered by decreasing size. This makes it easier to interpret the graph. Ordering by size should only be considered if the categories have no natural order.

Many bar charts in research reports and in the media have vertical bars instead of horizontal bars. This can lead to graphs that are less easy to read. See Figure 11.4 for an example. As there is not enough room for the labels, they have been rotated. The same is the case for the title of the vertical axis. Note that this bar chart resembles a histogram more closely, which may lead to some confusion.

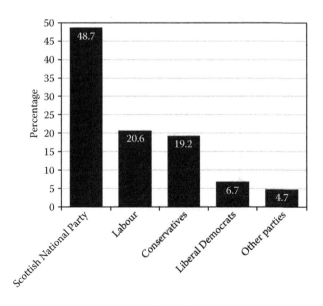

FIGURE 11.4 A bar chart with vertical bars. (From Survation, May 2016.)

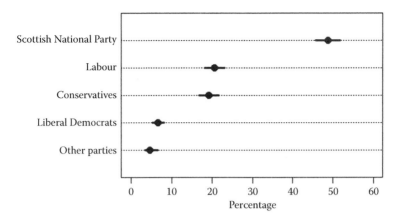

FIGURE 11.5   Dot plot for a Scottish poll. (From Survation, May 2016.)

## 11.3.3 Dot Plot

A less popular way to display the answers to a closed question is the dot plot. The dot plot was proposed in 1984 by William Cleveland, see Cleveland (1984). He conducted extensive research on the easiest ways to interpret graphs. One of his conclusions was that the dot plot does a better job than the bar chart.

Figure 11.5 shows a dot plot for voting intentions in the poll for the Scottish elections of May 2016. There is a horizontal dotted line for each answer option. The position of the big dots represents the size of the category. It is clear that the SNP is the largest party with approximately 50% of the votes.

An attractive property of a dot plot is that it is very easy to visualize the margins of error visible. They are represented by the short horizontal line segments in Figure. Taking these margins of error into account, there is indeed a large significant gap between the SNP and the other parties. As the margins of error of Labour and the Conservatives overlap, one must conclude that there is no significant difference between these two parties. The difference in the plot should be attributed to sampling noise, and not to real difference between the parties. It is *too close to call*.

## 11.3.4 Grouped and Stacked Bar Chart

The graphs described up until now show the answers to a single question. Often, however, one also wants to show possible relationships between two questions. This could provide insight in combinations of answers that are more frequent than other answer combinations. For showing the

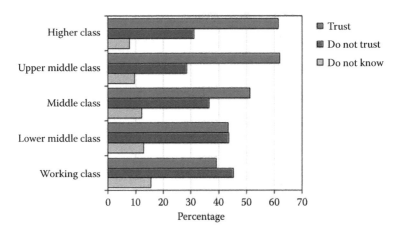

FIGURE 11.6   Grouped bar chart of social class by trust in the European Union. (From Eurobarometer.)

relationship between two closed questions, two types of graphs are very popular: the grouped bar chart and the stacked bar chart.

To show both types of bar charts, data are used that were collected in the Eurobarometer. The Eurobarometer is an opinion poll that is conducted regularly in all member states of the European Union. The European Commission is responsible for conducting the Eurobarometer. It is conducted twice a year. Among the questions asked are the assessment of the social class and the trust that one tends to have in the European Union. The data in this example originate from the spring version of Eurobarometer in 2015.

Figure 11.6 shows an example of a grouped bar chart. For each category of one variable (social class), there is a bar chart of the other variable (trust in the European Union). Note that there is some spacing between the bar charts, and there is no spacing between the bars within a category.

A clear trend can be observed in this bar chart. Trust in the European Union increases with social class. Higher class people are much more positive about the EU than working class people. Moreover, the percentage of people without an opinion decreases with increasing social class.

The stacked bar chart provides a different view on the relationship between two closed questions. The difference with a grouped bar chart is that the bars of the bar charts are now stacked onto each other. See Figure 11.7 for an example. This graph gives a better view on the composition of a variable within the categories of the other variable. Note, however, that it is now more difficult to see trends in the middle categories of the other variable.

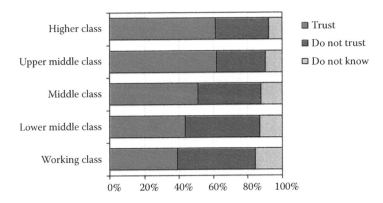

FIGURE 11.7   Stacked bar chart of social class by trust in the European Union. (From Eurobarometer.)

Note that both types of bar charts can be useful. So it is a good idea to take a look at both charts. Note also that the role of the two variables can be interchanged. In this case, this would mean making a graph of the distribution of the social classes within the categories of the trust variable. This could lead to other interesting conclusions.

### 11.3.5  Showing Developments over Time

If a poll is repeatedly conducted (e.g., every month, or every year), changes over time can be explored. This is typically the case for political polls, in which the researcher tries to make clear how party support changes in the course of time. There are two types of graphs that can be used: a bar chart and a line chart.

Figure 11.8 shows a bar chart. When showing developments over time, it is common to use the horizontal axes for time. That is why there are vertical bars here instead of horizontal bars. It is clear from the graph that trust in EU decreases until 2013, and from that year it rises again.

It is appropriate to make a bar chart if a variable is measured at discrete points in time. The bars are well suited to make comparisons between these time points. Note there is some spacing between the bars to avoid confusion with other types of graphs (like a histogram). Note also that the distance between the bars should reflect the amount of time that has passed between the corresponding time points. For example, if there was no measurement for the year 2013, the distance between the bars for 2012 and 2014 would not change, resulting in a wide gap between the two bars.

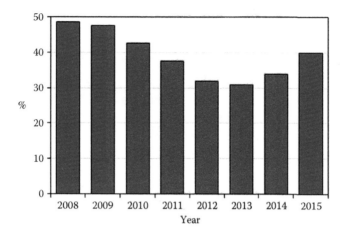

FIGURE 11.8   Bar chart showing trust in the European Union over time. (From Eurobarometer.)

If a variable continuously changes over time, and focus is not on levels but on changes, it is better to use a line chart. Figure 11.9 shows and example of such a chart. For the blue line, the same data were used as in Figure 11.8.

The time point at which polls were conducted are not shown in the chart. It is possible to add these time points by drawing small points, squares, circles, or other symbols. One should realize that adding these points often does not help the interpretation of the chart.

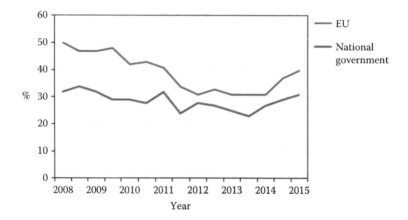

FIGURE 11.9   Line chart showing trust over time in the European Union and in the national government. (From Eurobarometer.)

It is one of the advantages of line graphs that development of several variables can be shown simultaneously, by including a line for each variable. Note, however, that chart should not have too many lines, as this would obscure its interpretation.

A second line has been added to the chart in Figure 11.7. This shows the trust of respondents in their national government. The message of the chart is clear: trust in the European Union is systematically higher than trust in the own national government.

A legend has been included to be able to distinguish the various lines. Note that the order of the lines in the legend is the same as the order of the lines in the graph. This helps us to locate lines.

## 11.4 GUIDELINES FOR DESIGNING GRAPHS

A graph can be a powerful tool to convey a message contained in a poll data set, particularly for those without knowledge of statistics. Graphs can be more meaningful and more attractive than tables with numbers. Not surprisingly, graphs are often used in the popular media like news websites, newspapers, and television. Use of graphs is, however, not without problems. Poorly designed graphs may convey the wrong message. There are ample examples of such graphs. Designers without much statistical knowledge often make them. They pay more attention to attractiveness of the graphical design than to its statistical content.

To avoid problems with graphs, a number of design principles should be followed. Some of these guidelines are discussed in this section. Also some examples are given of badly designed graphs. More about good and bad graphs can, for example, be found in Tufte (1983), Wainer (1997), Robbins (2013), and Bethlehem (2015).

### 11.4.1 Guideline 1: Do Not Mess with the Scales

The scales on the axes should help the reader to correctly interpret the magnitude of the displayed phenomena. If the measurement scale of a variable has a clear interpretation of the value 0, the axis should start at this value, and not at an arbitrary larger value, as this could lead to a wrong interpretation of the graph.

Figure 11.10 shows an example. The bar chart in Figure 11.10a shows the essential part of a chart published by *Fox News* in March, 2014. It intended to show the state of affairs with respect to Obamacare enrollment. Obamacare was the new health insurance system introduced by president Obama in the United States. According to government estimates,

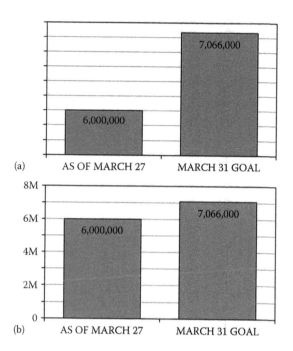

FIGURE 11.10 (a, b) Two bar charts showing enrollment of Obamacare. (From Fox News, New York, March 2014.)

7066,000 people should be enrolled by March 31, 2014. At first sight, the bar chart seems to suggest that by March, 27, enrollment is still far-off the target. Unfortunately, the vertical axis is missing. This makes interpretation difficult. In fact, 6000,000 people were enrolled by March, 27, which means the difference with the target is not very large. Apparently, the vertical axis did not start at 0, and this caused the difference between the two bars to be exaggerated. After a lot of criticism, *Fox News* published a new version of the graph a few days later. This is the bar chart in Figure 11.10b, which shows the essential part of the repaired bar chart. It has a properly labeled vertical axis, and this axis starts at 0. This bar chart gives a more realistic picture of the situation.

Figure 11.11 shows another example of incorrect design of axes. The line chart shown in Figure 11.11a was published (in Dutch) by *Statistics Netherlands* (*Webmagazine*, January 17, 2008), the national statistical institute of The Netherlands. It shows the increase of the average length of male and female adults in the years from 1982 to 2006. Some spectacular patterns can be distinguished. Apparently, the average length of

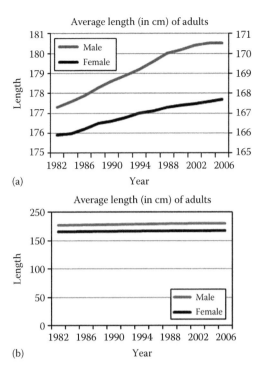

FIGURE 11.11 (a, b) Average lengths of males and females in The Netherlands. (From Netherlands Web magazine, January 17, 2008.)

both males and females is rapidly rising. Also, males are much longer than females. And the difference is getting larger.

A closer look at this line chart reveals two defects. The first one is that the graph has two vertical axes: one on the left for the length of males, and one on the right for the length of females. Note that the right-hand axis has shifted upward with respect to the left-hand axes. Therefore, both lines are not comparable. The second defect is that both axes do not start at 0. This makes differences more spectacular than they really are.

The defects were repaired in the line chart in Figure 11.11b. There is only one vertical axis, and the scale at this axis starts at 0. The rise in length is now far from spectacular. There is only a very modest rise. And the difference between males and females is small.

## 11.4.2 Guideline 2: Put the Data in the Proper Context

The graph should show the statistical message in the proper context, so that the right conclusion is drawn by the user. Often, graphs show only one question/variable at the time. Without having information about

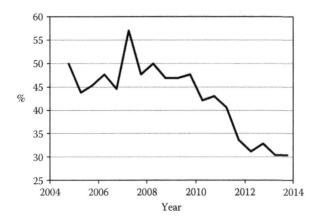

FIGURE 11.12    A line chart of trust in the European Union. (From VRT, Belgium, January 2014.)

the other variables, it may be hard to interpret the message. Design and composition of the graph should be such that the correct message is conveyed. This may mean to show more than just one variable. Of course, there should not be too many variables in the graph as this may obscure the message.

Figure 11.12 shows an example of a graph that lacks proper context. The original graph was shown at January 23, 2014 in the news program *Terzake* of the Flemish broadcasting company VRT. It is supposed to show the development of trust in the member states of European Union. The data were collected in a Eurobarometer poll.

At first sight, there seems to be substantial downward trend. The line even approaches the horizontal axis. So one might get the impression that there is almost no trust left. There are, however, two issues with this graph. The first issue is that the vertical axis does not start at 0% but at 25%. This amplifies the effects in the graph. The drop in trust looks more dramatic than it really is. The second issue is that one may wonder what it means that 31% of the people trust the European Union. Is this high, or is this low? There is no reference point.

Both issues are addressed in Figure 11.13. The vertical axis now starts at 0%. There still is a downward trend, but it is less dramatic. To provide more context, another variable has been added to the graph. It measures trust in the national government. From this graph, one can conclude that trust in the European Union is not so bad, as it is higher than the trust in the national government. It is also interesting that both lines have the same

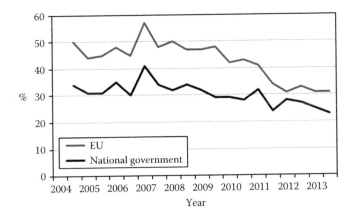

FIGURE 11.13 A line chart of trust in the European Union and the national government.

pattern of going up and down. This suggests that there are other reasons for the downward trend than the European Union itself.

Figure 11.14 shows another example in which lack of context leads to a wrong conclusion. It shows part of a billboard on a street in the town of Leeds, United Kingdom. The billboard was made by the British rail industry to show how satisfied British travelers are with rail services. Indeed, the United Kingdom seems to have the highest score with 78%.

This graph has at least three defects. The first one is that the bar chart has vertical bars (instead of horizontal ones) and no spacing between the bars. This may lead to confusion with other types of graphs.

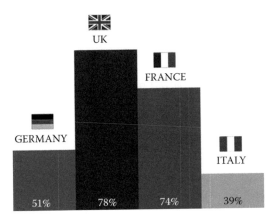

FIGURE 11.14 A bar chart of satisfaction with rail service in Europe. (Courtesy of James Ball.)

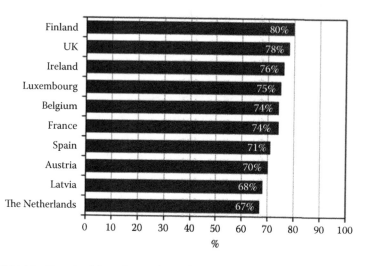

FIGURE 11.15   Repaired bar chart of satisfaction with rail service in Europe.

The second defect is that there seems to be something wrong with the lengths of the bars. For example, satisfaction in the United Kingdom is 78%, which is twice the satisfaction in Italy (39%). Consequently, the bar for the United Kingdom must be twice as long as the bar for Italy, but this is not the case. Probably, the vertical scale does not start at 0. Unfortunately, one cannot establish what is wrong with the vertical scale, because it is completely missing.

The third defect is that a large part of the context is missing. The graph only contains data about four countries. What about other countries in Europe? It turns out that these data were collected as part of a Eurobarometer poll in 2013 about satisfaction with rail services in Europe. Figure 11.15 contains the repaired barchart for the top 10 countries. The bars are now horizontal, they have spacing, and there is a (horizontal) scale that starts at 0. Now it is clear that the United Kingdom is not the best country from the point of view of satisfaction with rail services. Finland is doing better. Germany and Italy have disappeared from the top 10 because of their low scores.

Note that percentages are close together in the bar chart. Given that the sample sizes per country in the Eurobarometer are around 1000, differences are not significant. So one should be careful with statements about which country is better or worse than which other country.

### 11.4.3 Guideline 3: Be Careful with the Use of Symbols

Graphs are used to visually display the magnitude of phenomena. There are many techniques to do this. Examples are bars of the proper length, lines at the certain distance from an axis, or points on a scale. Whatever visual metaphor (symbol) is used to represent the magnitude, it must be such that it enables correct interpretation. For example, it should retain the natural order of the values. If a value is twice as large as another value, the user should interpret the metaphor of the first as twice as large as the second metaphor. Unfortunately, this is not always the case. Particularly graphs in popular printed media tent to violate this principle.

Figure 11.16 contains an example of a graph in which the symbols do not have the correct size. This bar chart is supposed to show what percentage of internet users in The Netherlands use a smartphone for this.

The designer of this graph decided using symbols instead of plain bars. As the graph is about smartphones, it seems reasonable to take smartphones as symbols. So the designer made the heights of the smartphones proportional to the percentage of smartphone use. This is a design error. If a smartphone is higher, it is also wider. When comparing different years, readers will look at the area that the symbol takes, and not at its height. So they interpret the graph incorrectly.

Tufte (1983) introduced the *lie factor* for this type of problems in graphs. It is defined as the value suggested by the graph divided by the true value. Ideally, the lie factor is close to 1. According to Figure 11.14, smartphone use has risen by a factor $59/22 = 2.7$ from 2008 to 2012. The areas of the

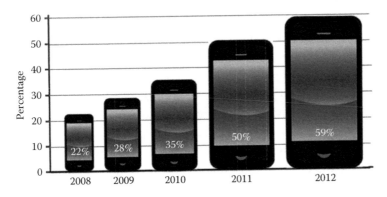

FIGURE 11.16   Percentage of internet users in The Netherlands using a smartphone for this. (From Statistics Netherlands, 2016.)

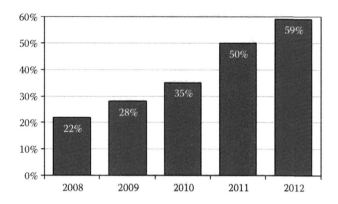

FIGURE 11.17 Percentage of internet users in The Netherlands using a smartphone for this (repaired version). (From Statistics Netherlands, 2016.)

smartphones have increased by a factor 7.2 in the same period. So the lie factor is here equal to $7.2/2.7 = 2.7$.

The problems in the symbol graph could be solved by taking the *areas* for the smartphone pictures proportional to the values of the variable (instead of their lengths). It may be an even better solution to avoid symbols and make a simple bar chart as in Figure 11.17.

Problems with the use of symbols in graphs exist already for a long time. In 1924, Escher (stepbrother of the famous Dutch graphic artist M.C. Escher) wrote a little book called *De methodes der grafische voorstelling*. See Escher (1924). He discussed the graph in Figure 11.18a, which shows how the number of schoolboys in The Netherlands increased from 1875 and 1915. He concluded that it was a bad graph. Indeed, the number of schoolboys increased, but not as fast as the graph suggests. The symbol for 1915 is 3.3 as large as the symbol for 1875, whereas the number of schoolboys went up by a factor 1.9. This is a lie factor of 1.7. Another problem of this graph mentioned by Escher was that some people concluded from the graph that schoolboys grew taller over the years.

There is a better way of working with symbols. Instead of trying to make the size of a symbol proportional to the corresponding value, one can keep the size of the symbols the same, but repeat the symbol a number of times, where the number of symbols reflects the value. So, if a value is twice as large, there are twice as many symbols. The graph in Figure 11.18b shows an example. It uses the same data as the graph shown in Figure 11.18a. Each symbol represents 20,000 schoolboys.

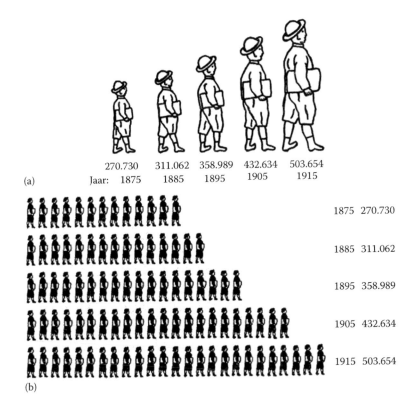

FIGURE 11.18    (a) The rise of the number of schoolboys in The Netherlands from 1875 to 1915 and (b) an example that uses the same data as (a). (Printed with permission of Nieuw Amsterdam from Escher 1924.)

## 11.4.4 Guideline 4: No Three-Dimensional Perspective

Often, graphs for the popular media are made by graphical designers instead of statisticians. Designers may find simple graphs boring and therefore may attempt to make them more attractive, for example, by adding chart junk. Another way to make a graph more attractive is to add a three-dimensional perspective to it. Many statistical packages (e.g., Microsoft Excel) support this. However, three-dimensional graphs are not a good idea from a statistical point of view, because they tend to make correct interpretation more difficult.

Figure 11.19 contains an example of a graph with a three-dimensional perspective. It is the three-dimensional version of the pie chart in Figure 11.1. The data are from a Scottish election poll, conducted by *Survation* in May 2016. The percentages for the Conservatives (19.2%) and Labour (20.6%) are almost the same. Nevertheless the (red) slice for Labour is much larger than the (blue) slice for the Conservatives. This is

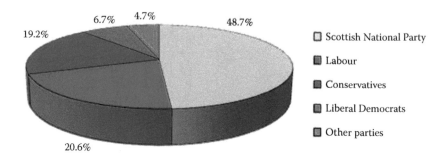

FIGURE 11.19 Three-dimensional pie chart for a Scottish election poll. (From Survation, May 2016.)

partly caused by the different distortion of the slices. This causes slices in the front to look larger than slices at the sides.

Moreover, due to the three-dimensional nature of the graph, the height of the pie becomes visible, but only for slices in front of the pie, and not for other slices. This also increases the impact of the (red) slide for Labour.

Also other types of graphs sometimes get a three-dimensional perspective. Figure 11.20 contains a three-dimensional version of bar chart. This bar chart is supposed to show the result of an opinion poll in The Netherlands. Similar versions of the bar chart were published by Dutch newspapers. The poll was conducted in October 2013 by *Peil.nl*. The blue pillars represent the results of the poll in seats in parliament. The blue pillars show the current composition of parliament.

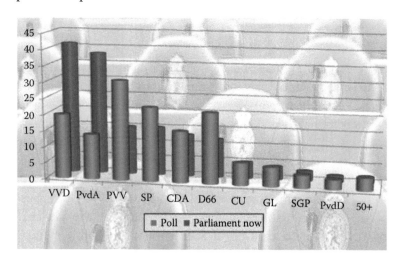

FIGURE 11.20 Three-dimensional bar chart for a Dutch political poll. (From Peil.nl, October 2013.)

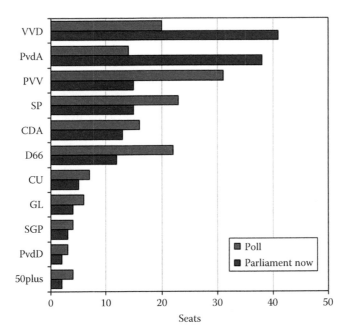

FIGURE 11.21 Grouped bar chart for a Dutch political poll. (From Peil.nl, October 2013.)

It is hard to interpret this graph. It is almost impossible to establish the correct lengths of the pillars. As an exercise, one could attempt to determine the length of the pillars for the PVV in the poll.

Another problem with this three-dimensional graph is that sometimes pillars are partly or completely hidden behind other pillars. For example, the numbers of seats in parliament of the small parties are almost invisible.

The problems of this three-dimensional graph can be solved by replacing it by a (two-dimensional) grouped bar chart. The result is shown in Figure 11.21. It is now easy to determine the lengths of the various bars, to compare bars for different parties, and to see differences between the poll and the current numbers of seats in the parliament.

## 11.4.5 Guideline 5: No Chart Junk

Particularly graphs in popular media may contain chart junk. *Chart junk* refers to all visual elements in graphs that are not necessary for correct interpretation of the message conveyed by the graph. To the contrary, chart junk may distract the viewer of the graph from this message. It serves no other purpose than making the picture more attractive from an artistic point of view. The term chart junk was invented by Tufte (1983). It is clear that chart junk must be avoided.

Tufte (1983) introduced the data-ink ratio (DIR) as a measure of the amount of chart junk in a graph. It is defined as the ratio of the amount of ink used to draw the nonredundant parts of the graph (the data, labels, tick marks, and values) and the total amount of ink used. An ideal graph would have a DIR value close to 1. Values of the DIR much smaller than 1 indicate that the graph contains too much chart junk.

The graph in Figure 11.22a is an example of a graph with chartjunk. The original version of the graph was published in the Washington Post

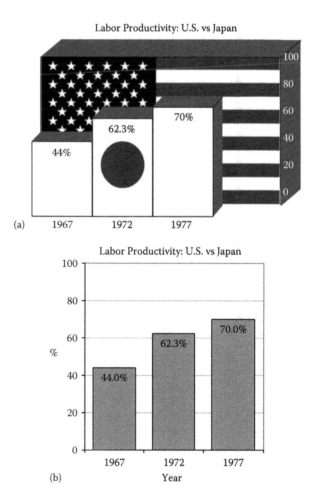

FIGURE 11.22 Labor productivity, (a) with and (b) without char junk. (From Washington Post, 1979.)

in 1978. It compares labor productivity of the United States and Japan for three points in time. The graph contains a lot of decoration that does not really help us to convey the statistical message. To the contrary, it obscures the message. Note that the graph only contains three numbers that must be shown: 44.0, 62.3, and 70.0. The DIR is clearly less than 1.

The chart junk has been removed in the graph shown in Figure 11.22b. It is just a simple bar chart. There is now a much more pronounced role for the data itself. The DIR of this graph will be close to 1.

Another example of chart junk can be found in the graph in Figure 11.20. The graph compares the estimates for the number of seats in parliament of a Dutch poll with the current number of seats. The graph is already hard to read because of its three-dimensional nature. What makes it even more difficult is the background picture. This is a photo of the chairs in the parliament building. This decoration does not help at all one to understand the message in the data. Removing the background will make the bar chart much more readable.

## 11.5  SUMMARY

A poll should ultimately produce some kind of research report. For a small poll, only a few pages may be sufficient. But larger polls may require a substantial report. Research reports can be structured in various ways, but it should at least contain three parts:

- *An executive summary*: This is a short description of the poll in a manner that is readable and understandable for everyone. It should describe the research question and give an overview of the most important conclusions.

- *A methodological account*: The methodological account is an accurate description of the how the poll was designed and carried out. This account should provide sufficient information to determine whether the conclusions were drawn in a scientifically sound way.

- *The outcomes of the analysis*: The analysis of the collected data could start with an exploratory analysis of the data. This exploratory analysis should provide insight in each variable (target variable or auxiliary variable) separately. This may be followed by a more in-depth analysis.

The results of a poll are often presented by means of graphs. A graph can be powerful instrument for conveying the message contained in the data, particularly for those without knowledge of statistics. Use of graphs is, however, not free from problems. Poorly designed graphs may easily convey the wrong message. There are ample examples of such bad graphs. Problems are often caused by designers of graphs who lack sufficient statistical expertise. Unfortunately, they pay more attention to attractiveness of the graphic design than to its statistical content.

This chapter describes how to make good graphs, and how to distinguish good graphs from bad graphs. Some guidelines may help in this respect. Several examples of bad graphs in the media show that one must always be careful when viewing graphs.

# CHAPTER 12

# A Checklist for Polls

## 12.1 SEPARATING THE CHAFF FROM THE WHEAT

Many polls are conducted everywhere in the world. Particularly, election periods show an increase in the number of opinion polls. There are ample examples of countries in which multiple election polls are carried out almost every day of the election campaign. And even outside election periods, there are more and more polls. Opinions of citizens are asked about an ever wider variety of topics.

The increase in the number of polls is mainly caused by the rapid developed of the internet. Internet makes it possible to conduct online polls, and this is a fast, easy, and cheap way to collect a lot of data. There are websites (e.g., www.surveymonkey.com) in which everyone, even people without any knowledge of research methods, can set up a poll very quickly. The question is, however, whether these polls are good polls. If not, the validity of its outcomes is a stake.

All polls have in common that a sample of people is selected from a target population. And all people in this sample are asked to complete a questionnaire. The questions can be about facts, behavior, opinions, and attitudes. The researcher analyzes the sample data and draws conclusions about the population as a whole. That can be done in a meaningful way, provided the poll has been set up and conducted according to scientifically sound principles.

So, if a press release or publication about a poll is encountered, first the question must be asked and answered whether the poll is a good one and hence, its outcomes are valid, or that the poll is a bad one and that it is probably better to ignore its outcomes. It can be difficult for users of poll results (journalists, policy-makers, and decision-makers) to distinguish the chaff from the wheat.

To help them, a checklist has been developed. By going through its nine questions one by one, an impression can be obtained of the quality and the usefulness of the poll. If the quality looks good, more attention can be paid to the outcomes of the poll. If the checklist raises concern about the quality of the poll (too many checklist questions were answered by *no*), or a lot of information about the poll is missing, it may be better to ignore its outcomes.

The questions in the checklist cannot be answered if there is no proper poll documentation available. Such a research report must allow its reader to get a good idea about how the poll was set up, and how it was carried out. So what kind of methodological information about the poll is needed? This was already described in detail in Chapter 10. This list is based on several sources that describe what should be in the research report. Among these sources are the National Council on Public Polls, the American Association of Public Opinion Research, the World Association for Public Opinion Research, and ESOMAR. The checklist in this book is directly based on the checklist for polls developed by the Dutch-speaking Platform for Survey Research Nederlandstalig Platform voor Survey-onderzoek (NPSO) together with Statistics Netherlands Centraal Bureau voor de Statistiek (CBS), and the Dutch-Flemish Association for Investigative Journalism Vereniging van Onderzoeksjournalisten (VVOJ) (see Bethlehem, 2012).

The checklist consists of the nine questions in Table 12.1. The subsequent section explains in more detail why these questions are so important.

TABLE 12.1  The Checklist for Polls

| | Question | Yes | No |
|---|---|---|---|
| 1 | Is there a research report explaining how the poll was set up and carried out? | | |
| 2 | Is the poll commissioned or sponsored by an organization that has no interest in its outcomes? | | |
| 3 | Is the target population of the poll clearly defined? | | |
| 4 | Is a copy of the questionnaire included in the research report, or elsewhere available? | | |
| 5 | Is the sample a random sample for which each person in the target population has a positive probability of selection? | | |
| 6 | Are the initial (gross) sample size and the realized (net) sample size (number of respondents) reported? | | |
| 7 | Is the response rate sufficiently high, say higher than 50%? | | |
| 8 | Have the outcomes been corrected (by adjustment weighting) for selective nonresponse? | | |
| 9 | Are the margins of error specified? | | |

## 12.2 THE NINE QUESTIONS

### 12.2.1 Is There a Research Report?

There should be a report describing how the poll was set up and carried out. This report should contain sufficient information to assess whether the poll was done in a scientifically sound way. So the report should not only describe the outcomes of the poll but also the methodological aspects. The report should at least contain the following information:

- The organization that commissioned or sponsored the poll.

- The organization that conducted the poll.

- The definition of the target population. This is the group from which the sample was selected, and to which the conclusions of the poll relate.

- The questionnaire. It must also be clear whether the questionnaire was tested, and how it was tested.

- The sampling frame used to select the sample. It is the list with contact information (name, address, telephone number, or e-mail address) of every person in the target population.

- The way in which the sample was selected. It must be clear whether a random sample was drawn, and how the random sample was drawn. Was the sample drawn with equal or unequal probabilities? If the sample was selected from an online panel, how were its members recruited (random sample or self-selection)?

- The initial (gross) sample size. This is the size of the sample that was drawn from the sampling frame.

- The realized (net) sample size. This is the number of respondents.

- The response rate ($100 \times$ number of respondents/initial sample size).

- The way in which the poll was corrected for the effects of nonresponse (and possible other selection effects). If adjustment weighting was carried out, there should be a list of the auxiliary variables used. The report should make clear how the correction weights were computed.

- The margins of error. Note that these margins can only be computed if a random sample was selected, and nonresponse was not selective. In the case of a substantial amount of nonresponse, or self-selection, the researcher should warn that the differences between estimates, and true values can be larger than indicated by the margins of error.

### 12.2.2 Is the Poll Commissioned or Sponsored by an Organization That Has No Interest in Its Outcomes?

It is important to know who commissioned or sponsored a poll, because such an organization may have an interest in certain outcomes. It is not uncommon to see press releases about polls concluding that certain products or services are very good. A closer look often shows that the poll was conducted by companies offering these products or services. So, the poll is more a marketing instrument than a case of objective research.

One should be very careful if the poll is conducted by the organization that commissioned it. This concern caused the BBC (2010) to make an editorial guideline for this situation:

> If the research has been commissioned by an organization which has a partial interest in the subject matter, we should show extra caution, even when the methodology and the company carrying it out are familiar. The audience must be told when research has been commissioned by an interested party.

### 12.2.3 Is the Target Population Clearly Defined?

The target population is the group of people from which the sample was drawn and to which the outcomes of the poll refer. There must be a clear definition of the target population. It must always be possible to decide in practical situations whether or not a person belongs to the target population.

Problems may occur if the sampling frame does not cover the target population. In the case of undercoverage, the sample is selected from only part of the target population. Consequently, the outcomes only relate to this subpopulation and not to the whole population. Differences between the target population and the sampling frame must be reported.

For example, the target population was defined as all Dutch of age 18 years and older, but the sample was selected from only those with access to the internet. In fact, the conclusions only concern Dutch of age 18 years and older with access to internet.

### 12.2.4 Is the Questionnaire Available?

A good questionnaire is of vital importance. Indeed, practice has shown that it is easy to influence the outcomes of a poll by manipulating the texts of questions, and the order of the questions. It must be possible to determine the quality of the questionnaire. Therefore, the questionnaire must be available for inspection.

A good questionnaire contains objective and comprehensible questions. Here are a number of examples of pitfalls to avoid:

- *Incomprehensible questions*: Respondents may not understand questions because of the use of jargon, unfamiliar terms, or long, vague, and complex sentences. For example, *Are you satisfied with the recreational facilities in your neighborhood?*

- *Ambiguous questions*: For example, *When did you leave school?* What kind of answer is expected? A date, an age, or may be some event (when I married)?

- *Leading questions*: For example, *Most people feel that € 5 is way too much money to pay for a simple cup of coffee. Would you pay € 5 for a cup of coffee?*

- *Double questions* also called *double-barreled questions*: For example, *Do you think that people should eat less and exercise more?*

- *Negative questions* or *double-negative questions*: For example, *Would you rather not use a non-medicated shampoo?*

- *Recall questions*: Questions requiring recall of events that had happened in the past are a source of errors. People tend to forget these events. For example, *How many times did you contact your family doctor in the last two years?*

A good poll requires thorough testing of the questionnaire before it is used for data collection in *the field*. It should be clear how the questionnaire was tested.

### 12.2.5 Is the Sample a Random Sample?

To be able to draw meaningful conclusions about a target population, probability sampling must be applied. The sample must be a random sample. Everyone in the population must have a nonzero probability of being selected in the sample. The selection probabilities must be known.

The simplest way to select a random sample is to draw a *simple random sample*. This implies that everyone in the population has the same probability of selection. The analogy principle applies, which means that sample mean (or percentage) is a good estimator of the population mean (or percentage).

It is possible to select a sample with unequal probabilities. As a consequence, the estimators are somewhat more complicated, because they have to be corrected for the unequal probabilities. An example is an approach in which first addresses are drawn with equal probabilities, after which one person is drawn at random at each selected address. In this case, persons in large households have smaller selection probabilities than persons in small households.

If no probability sampling was applied, selection probabilities are unknown, making it impossible to compute valid estimates. This is, for example, the case for a quota sampling, and for self-selection sampling.

Many online polls are based on a random sample from an online panel. Then it is important know how the panel members were recruited. If the panel is a random sample, a random sample from this panel is also a random sample. If the online panel is based on some other form of sampling such as self-selection, it is not representative. Consequently, a random sample from this panel is not a random sample from the population.

### 12.2.6 Are the Initial Sample Size and Realized Sample Size Reported?

The initial sample size (gross sample size) and the realized sample size (net sample size) tell you something about the validity and the precision of the estimates. In the ideal case, the sample is selected by means of probability sampling, and there is no nonresponse. All people in the sample have filled in the questionnaire. So the realized sample is equal to initial sample size.

There is a simple rule in the ideal case: the precision of the estimates increases as the sample size increases. The precision is usually indicated by the margin of error. The margin of error is the maximum difference between the estimate and the true population value. Table 12.2 contains the margin of error for estimating a population percentage. The margin of error is large if the percentage is close to 50%. The margins of error decrease as the sample size increases.

The use of the table is illustrated with an example. Suppose a poll was conducted with a simple random sample of size 500. It turns out that 40%

TABLE 12.2   Margins of Error

| Percentage | Sample Size | | | | | |
|---|---|---|---|---|---|---|
| | 100 | 200 | 500 | 1000 | 2000 | 5000 |
| 10 | 5.9 | 4.2 | 2.6 | 1.9 | 1.3 | 0.8 |
| 20 | 7.9 | 5.6 | 3.5 | 2.5 | 1.8 | 1.1 |
| 30 | 9.0 | 6.4 | 4.0 | 2.8 | 2.0 | 1.3 |
| 40 | 9.7 | 6.8 | 4.3 | 3.0 | 2.1 | 1.4 |
| 50 | 9.8 | 6.9 | 4.4 | 3.1 | 2.2 | 1.4 |
| 60 | 9.7 | 6.8 | 4.3 | 3.0 | 2.1 | 1.4 |
| 70 | 9.0 | 6.4 | 4.0 | 2.8 | 2.0 | 1.3 |
| 80 | 7.9 | 5.6 | 3.5 | 2.5 | 1.8 | 1.1 |
| 90 | 5.9 | 4.2 | 2.6 | 1.9 | 1.3 | 0.8 |

of the respondents are in favor of a certain government policy. Table 12.2 shows that the margin of error for a sample size of 500 and an estimate of 40% is equal to 4.3%. So, the percentage in favor of the government policy in the target population is (with a high probability) somewhere between $40 - 4.3 = 35.7\%$ and $40 + 4.3 = 44.3\%$.

Suppose, an election poll was carried out with a sample size of 1000 people. The result was that 20% of the respondents expressed the intention to vote for a certain party. One month later the poll is carried out again. This time 22% of the respondents indicate they will vote for the party. Can one conclude that support for the party has increased? No, because both percentages have a margin of error of 2.5%. The margin of error is larger than the difference between the two percentages $(22 - 20 = 2\%)$. So the difference between the two polls can be attributed the sample *noise*.

If there was nonresponse in the poll, the realized sample size is smaller than the initial sample size. This implies that estimates not only have a margin of error, but they can also have a bias due to nonresponse. The accuracy of estimates has two components: precision and bias. The precision can be controlled. A large realized sample means a high precision. Unfortunately, the bias cannot be controlled. It is unknown, and increasing the sample size does not help.

## 12.2.7  Is the Response Rate Sufficiently High, Say Higher than 50%?

If there is nonresponse in a poll (the realized sample size is smaller than the initial sample size), it is important to know the response rate.

The response rate is obtained by dividing the realized sample size by the initial sample size. Often, the response rate is expressed as a percentage:

$$\text{Response rate} = 100 \times \frac{\text{Realized sample size}}{\text{Initial sample size}}$$

The lower the response rate, the more serious the nonresponse problems are. Which response rate is acceptable, and which one is not? Maybe the rule of thumb below will help, but realize that the bias of estimates does not only depend on the response rate (see Chapter 7).

Do not worry if the response rate is 90% or higher, as nonresponse will have little impact. A response rate of around 60% is still reasonable, but it is good to apply some adjustment weighting in order to correct for the effects of nonresponse. Worry if the response rate is below 50%. The estimates will probably have a substantial bias. Heavy adjustment weighting is called for, but there is no guarantee that this solves all nonresponse problems. Finally, a response rate of below 30% means serious problems. The quality of the poll is so affected that it is almost impossible to compute valid estimates.

Usually it is impossible to determine the magnitude of the nonresponse bias. One can only do that, if the answers of the nonrespondents can be compared with those of the respondents. But the nonrespondents did not answer these questions. What to do? One thing that can be down is computing the worst case: how large can the bias at most be?

Suppose, an election poll was carried out, and the response rate was 40%. Of the respondents, 55% said they will vote at the election. If 40% responded, 60% did not respond. There are two extreme situations. The first one is that all nonrespondents will not vote. Then, the percentage of voters in the complete sample would be equal to

$$0.40 \times 55\% + 0.60 \times 0\% = 22\%$$

The second extreme situation is that all nonrespondents will vote. Then, the percentage of voters in the complete sample would be equal to

$$0.40 \times 55\% + 0.60 \times 100\% = 82\%$$

So the estimate for the percentage of voters (55%) could also have been any other percentage between 22% and 82%. The bandwidth is large. Indeed, the effects of nonresponse can be substantial. In fact, one must conclude the outcomes of this poll are not meaningful.

TABLE 12.3    Bandwidth of the Estimator Due to Nonresponse

| | Response Rate | | | |
|---|---|---|---|---|
| Percentage in Response | 20 | 40 | 60 | 80 |
| 10 | 2–82 | 4–64 | 6–46 | 8–28 |
| 20 | 4–84 | 8–68 | 12–52 | 16–36 |
| 30 | 6–86 | 12–72 | 18–58 | 24–44 |
| 40 | 8–88 | 16–76 | 24–64 | 32–52 |
| 50 | 10–90 | 20–80 | 30–70 | 40–60 |
| 60 | 12–92 | 24–84 | 36–76 | 48–68 |
| 70 | 14–94 | 28–88 | 42–82 | 56–76 |
| 80 | 16–96 | 32–92 | 48–88 | 64–84 |
| 90 | 18–98 | 36–96 | 54–94 | 72–92 |

Table 12.3 contains the bandwidth of the complete sample percentages for a series of response rates. It is clear that the bandwidth decreases as the response rate increases.

## 12.2.8  Have the Outcomes Been Corrected for Selective Nonresponse?

It is important to correct the outcomes of a poll for the negative effects of nonresponse. Usually some kind of adjustment weighting is applied. This means that a weight is assigned to each respondent. These weights are computed in such a way that the response of the poll is corrected for over- and underrepresentation of specific groups.

An example: Suppose, a poll is carried out in the town of Rhinewood, and it turns out that the response consists for 60% of males and 40% of females. Then there is a discrepancy with the population distribution, because the population of Rhinewood consists for 51.1% of males, and for 48.9% of females. Apparently, males have responded better in the poll. They are overrepresented. To correct for this, every male respondent is assigned a correction weight equal to

$$\frac{51.1}{60.0} = 0.852$$

This means that every responding male will count for 0.852 instead of 1. The weight is smaller than 1, because there were too many males among the respondents. Each female is assigned a correction weight equal to

$$\frac{48.9}{40.0} = 1.223$$

So each female counts for 1.223 instead of 1. The correction weight is larger than 1, because there were too few females among the respondents.

By assigning these correction weights, the response becomes representative with respect to gender. The weights can be computed because the true population percentages are known. The idea of weighting is making the response representative with respect to as many variables as possible. However, only variables can be used that have been measured in the poll and for which the population distribution is available. This can be a serious restriction. Often used weighting variables are gender, age, marital status, and region of the country.

The hope is that if the response is made representative with respect to many auxiliary variables, the response also becomes representative with respect to the target variables of the poll. Not every adjustment weighting procedure is, however, effective. Weighting is only able to reduce the non-response bias if two conditions are satisfied:

- There is a strong relationship between the target variables of the poll and the auxiliary variables used for weighting.

- There is a strong relationship between response behavior and the auxiliary variables used for weighting.

For example, if an opinion poll is conducted, and the researcher decides to weight the response by gender and age, then this weighting procedure will be ineffective if there is no relationship between the variables gender and age and the variable measuring the opinion.

The researcher should always attempt to determine groups that are under- or overrepresented, what kind of effect this can have on the estimates, and whether adjustment can improve the situation.

The research report should contain information about weighting adjustment. Has a weighting adjustment procedure been carried out? If not, why not? And if so, which auxiliary variables were used, and how were they used. It would be informative to include a comparison of the response distribution and the population distribution of each auxiliary variable. That would make clear which groups are underrepresented, and which groups were overrepresented.

## 12.2.9 Are the Margins of Error Specified?

The outcomes of a poll are just estimates of population characteristics. Therefore, it is not realistic to assume the estimates are exactly equal to the true population values. Even in the ideal case of a simple random sample with full response, there still is a discrepancy between the estimate and the true, but unknown, population value. This is the *margin of error*. The researcher should compute the margins of error and include them in the research report.

An example: A simple random sample of 500 persons was selected. Everyone participates, so there is no nonresponse. Of the respondents, 60% say that they will vote at the next elections. The corresponding margin of error is 4.3%. Hence, the percentage of voters in the population will (with a high probability) be between $60 - 4.3 = 55.7\%$ and $60 + 4.3 = 64.3\%$.

Note again that the margins of error only indicate the uncertainty caused by sample selection. Nonresponse causes extra problems in the form of a bias. This bias is not included in the margin of error. So, in practice, the situation can be worse than indicated by the margin of error. The margin of error is only a lower bound. The real margin can be larger.

If a poll is repeated, one must realize that a small difference between polls does not have to indicate a *real* difference. It may just have been caused by sampling *noise*. Only if the differences are larger than the margins of error, can one conclude that something really changed. Note that the situation can be even more complicated if nonresponse occurs in both polls.

## 12.3 AN EXAMPLE: SOCIAL MEDIA STRESS

Use of the checklist is illustrated with an example of a Dutch poll. It was a poll about social media stress. There was a lot of interest for the outcomes of this poll in the Dutch media. These media reported the poll results but never asked themselves the question how good this poll was.

> Recent research of the National Academy for Media & Society shows that young people between the age of 13 and 18 years suffer from a serious form of Social Media Stress (SMS). Social media are getting a hold on young people with their subtle stimuli, such as sounds, push messages, attention, and rewards. Young people are unable to take an independent decision to stop, because they fear to be excluded. The serious form of this fear is called FOMO (Fear of Missing Out).

This was the first part of a press release published by the *National Academy for Media & Society* in May, 2012. Many news media took this up as a serious item. National radio and TV channels, national newspapers, and many news websites had stories about the dangers of social media.

What this media failed to do was checking the press release. The National Academy for Media & Society was an unknown organization. What kind of an organization was it? How did it reach the conclusions about social media stress? And how valid were these conclusions?

This section checks the facts about this poll. The checklist for polls is applied, and an attempt will be made to answer the nine questions.

### 12.3.1 Is There a Research Report?

It was possible to download a poll report from the internet. This report focused on the outcomes of the poll. There was very little about the way the poll was set up and conducted. The information was insufficient to assess the quality of the poll.

Figure 12.1 gives an idea of how the outcomes of the poll were presented in the report. The respondents were asked to what extent three statements applied to them (completely, somewhat, or not at all). The statements were (1) I feel stressed when I notice I cannot keep up-to-date with everything in the social media; (2) I have to process too much information in the social media to keep up-to-date; and (3) Because I want to keep up-to-date with everything, and I do not have time to immerse myself, I read information in the social media only superficially. Note that the three-dimensional nature of the graph does not really help to read it.

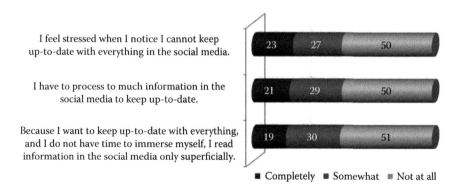

FIGURE 12.1  Some outcomes of the poll on Social Media Stress. (From Nationale Academie voor Media & Maatschappij 2012.)

The conclusions of the poll were not always supported by the collected data. For example, the conclusions of the report contain the statement that young people between the age of 13 and 18 years suffer from a serious form of social media stress. However, from charts like the one in Figure 12.1 one can conclude that at least half of the young people do not have a problem at all.

### 12.3.2 Is the Poll Commissioned or Sponsored by an Organization That Has No Interest in Its Outcomes?

The poll was an initiative of the National Academy for Media & Society. This organization also conducted the poll. Until the publication of the poll results, almost nobody knew it existed. The Academy turned out to be a small foundation, run by two people. It offered courses like *Social Media Professional* and *Media Coach*. So the organization had an interest in certain outcomes of the poll: the more problems the social media cause, the more attractive the courses become, and the more money the Academy could earn.

The conclusion was that the poll was not conducted by a neutral organization. Therefore, one should be very careful with the interpretation of the results.

### 12.3.3 Is the Target Population Clearly Defined?

There was some confusion about the target population. The press release seemed to suggest that the target population consisted of all young people between the ages of 13 and 18 years. However, the estimates in the report assumed the target population to consist of all young people between the ages of 13 and 18 years with a smartphone. So the conclusions of the poll are only related to young people with a smartphone and not to all young people.

### 12.3.4 Is the Questionnaire Included in the Research Report?

Unfortunately, the questionnaire was not included in the research report. Therefore, not much can be said about the quality of the questionnaire. It was also not clear whether the questionnaire was properly tested.

The titles of the graphs in the report seem to contain question texts. These texts raise some concern about the jargon used. What do the respondents exactly mean when they say the *feel stressed*? And what is the meaning of *keeping up-to-date with everything*? There is no guarantee that all respondents interpret these terms in the same way.

### 12.3.5 Is the Sample a Random Sample for Which Each Person in the Target Population Has a Positive Probability of Selection?

There was a problem with sample selection. The research report did not contain any information on the way the sample was selected. One could only read that 493 young people of an age between 13 and 18 years completed an online questionnaire.

After sending an e-mail to the Academy with a request for more information about sampling procedures, an e-mail was returned with some more details. Apparently, the National Academy for Media & Society had a network of 850 so-called youth professionals. These youth workers were active in education, libraries, youth counseling, and community work. A number of these youth workers were asked to invite approximately 20 young people to complete the questionnaire on the internet. In the end, 240 boys and 253 girls did so.

There are serious doubts about this sample design. No random sampling took place. The youth workers were not sampled at random from all youth workers. And young people were not sampled at random by the youth workers. Moreover, young people not involved in any kind of youth work could not be selected at all. So the conclusion is that this is not a representative sample. Therefore, it is unlikely the sample results can be generalized to the target population.

### 12.3.6 Are the Initial Sample Size and the Realized Sample Size (Number of Respondents) Reported?

The realized sample size was 493 young people. This is not a very large sample. So it is important to keep in mind that there are margins of errors. For example, if 50% of the young people in the sample say they not feel stressed, the percentage in the population is somewhere between 45.6% and 54.4%. This is based on a margin of error of 4.4% (see Table 12.2).

There is no information about the initial sample size. So it is not clear whether there was nonresponse in the poll. It is unlikely that all young people asked, indeed filled in the questionnaire.

So the sample is not a random sample, and there is an unknown amount of nonresponse. Hence, it is doubtful whether or not one can say something meaningful about the accuracy of the estimates. The only thing one can do is computing margins of error assuming a simple random sample, and hope and pray that there is no serious nonresponse bias. Anyway, one has to be extremely careful with the outcomes of this poll.

### 12.3.7 Is the Response Rate Sufficiently High, Say Higher than 50%?

It was already mentioned that there is nothing about nonresponse in the research report. One may expect, however, that there are youth professionals and young people who do not want to cooperate. If those people differ from the people that want to participate, there may be a serious bias.

It is very unfortunate that nothing has been reported about nonresponse and the problems it may cause. This may give a too optimistic picture of the poll.

### 12.3.8 Have the Outcomes Been Corrected (By Adjustment Weighting) for Selective Nonresponse?

Although nothing was said about nonresponse, the researchers indicate they applied a *mild* form of adjustment weighting. They corrected their sample by comparing the distribution of males and females in the response by that of males and females in the population. Unfortunately, this had no effect. They constructed their sample such that it more or less had the proper numbers of males and females. So adjustment weighting could not improve the situation. The sample was already representative with respect to gender.

### 12.3.9 Are the Margins of Error Specified?

The research report of the poll on social media stress did not contain margins of error, or any indication that figures in the report were just estimates.

For percentages, the users of the poll results can compute the margins of errors themselves. This is only possible if a random sample was selected and nonresponse is not selective. The expression for the (estimated) margin of error of a sample percentage $p$ is equal to

$$M = 1.96 \times \sqrt{\frac{p \times (100 - p)}{n}}$$

To compute the margin of error for the percentage of people suffering from social media stress, substitute $p = 23$ (see Figure 12.2, first bar), and $n = 493$. The margin of error becomes

$$M = 1.96 \times \sqrt{\frac{23 \times 77}{493}} = 3.7$$

So the population percentage will be (with a high probability) between $23 - 3.7 = 19.3\%$ and $23 + 3.7 = 26.7\%$.

## 12.4 SUMMARY

The poll on social media stress was carried out by an organization with an interest in social media stress problems. There was no copy of the questionnaire. No proper sample was selected. Nonresponse was not reported and not corrected. The conclusions were not always supported by the data. As becomes clear from the completed checklist in Table 12.4, a lot was wrong with this poll. All-in-all, one should not attach too much importance to this poll. It is better to ignore it.

TABLE 12.4  The Checklist for Polls Applied to the Poll about Social Media Stress

|   | Question | Yes | No |
|---|----------|-----|-----|
| 1 | Is there a research report explaining how the poll was set up and carried out? | | √ |
| 2 | Is the poll commissioned or sponsored by an organization that has no interest in its outcomes? | | √ |
| 3 | Is the target population of the poll clearly defined? | √ | |
| 4 | Is a copy of the questionnaire included in the research report? | | √ |
| 5 | Is the sample a random sample for which each person in the target population has a positive probability of selection? | | √ |
| 6 | Are the initial (gross) sample size and the realized (net) sample size (number of respondents) reported? | | √ |
| 7 | Is the response rate sufficiently high, say higher than 50%? | | √ |
| 8 | Have the outcomes been corrected (by adjustment weighting) for selective nonresponse? | | √ |
| 9 | Are the margins of error specified? | | √ |

# Index

Note: Page numbers followed by f and t refer to figures and tables, respectively.